Computer-Aided Manufacture in Architecture

Computer-Aided Manufacture in Architecture

The Pursuit of Novelty

Nick Callicott

Architectural Press

OXFORD   AUCKLAND   BOSTON   JOHANNESBURG   MELBOURNE   NEW DELHI

Architectural Press
An imprint of Butterworth-Heinemann
Linacre House, Jordan Hill, Oxford OX2 8DP
225 Wildwood Avenue, Woburn, MA 01801-2041
A division of Reed Educational and Professional Publishing Ltd

 A member of the Reed Elsevier plc group

First published 2001

British Library Cataloguing in Publication Data
Callicott, Nick
   Computer-aided manufacture in architecture: the pursuit of novelty
   1. Computer-aided design   2. Architectural design – Data
   processing
   I. Title
   721'.0285

Library of Congress Cataloguing in Publication Data
A catalogue record for this book is available from the Library of Congress

ISBN 0 7506 4647 0

Composition by Scribe Design, Gillingham, Kent
Printed and bound in Great Britain
by Cromwell Press Ltd, Trowbridge

Contents

Acknowledgements

This book would have been an impossible task but for the help and contributions of many friends and colleagues who have shared their time, thoughts and skill.

Firstly, to my fellow partners in the research group Sixteen*(makers), Bob Sheil, Phil Ayres and Chris Leung who endured my absence from normal duties with great patience and understanding, but more importantly whose collaboration enabled the essential practical investigations within. Many of my colleagues at UCL are also part of this book: Abi Abdolwahabi, John Bremner, Bim Burton, Nat Chard, Stuart Dodd, Terry Jones, Dr Robin Richards and Alan Taylor have all, at various times, increased my experience and knowledge of all things made.

A vital contribution was made by Stephen Gage and Jeremy Till, whose careful reading of preliminary manuscripts provided a wealth of insight and constructive criticism, which I have acted upon whenever possible. In a similar vein I must also acknowledge a great debt to the late Steven Groák, his ideas form a large part of the inspiration and content of this book far more than can be expressed here.

Many firms and manufacturers have also freely contributed scarce time and resources. The Thermwood Corporation, Hardinge Machine Tools Ltd, DTM Corporation, 3D Systems Europe Ltd, Messer Cutting & Welding AG, MERO System GmbH & Co. and Charmilles Technologies have all kindly provided invaluable illustrations and technical information.

Special thanks are necessary to Jürgen and Brigitte Ehlert, Lothar Bühmann, Siegfried Pöhlke and everyone at the works of Ehlert GmbH whose limitless generosity, hospitality and humour made the 'Making Buildings' exhibit both a possibility and a pleasure to realise.

But most of all to Frank, Ruth and Kristina, who together dispelled the myth that writing is a lonely exercise.

Introduction

The technology of computer-aided manufacturing (CAM) has become an object of desire and mystery for many designers. For at least a decade we have enthusiastically ushered the computer into our studios and offices, commencing a revolution in the way we work and communicate our designs. Thoughtful practitioners have seen the computer at the heart of a distinctive and expressive medium, incomparable in many respects to previous methods of representation. Our fascination with CAM promises to extend this evolving understanding of representation still further, as we finally complete the cycle from conception to execution.

Currently, the debate regarding the immeasurable impact of the computer upon design practices remains as prolific as it is valuable yet, for all its insight, there is a sense that its richness and vitality is becoming diminished, not by a lack of imagination on the part of its authors, but simply by the increasing prevalence of technology's effects. Whilst the contribution to this theoretical debate has been substantial, a parallel exploration of this technology through 'project-based' design and research, particularly within architectural practice, appears to be less developed. In some instances this condition might reasonably be attributed to the difficulty of applying emerging developments, often still in their infancy, to genuine practical effect, but with CAM I believe this argument has recently become untenable.

Anecdotally, it is clear that the enthusiasm shown towards CAM by architects and other designers is more than matched by genuine confusion regarding its nature, capabilities and terms. This is quite understandable, not because of any complexity in the techniques themselves, but simply in the lack of familiarity many of us have with them. Few designers have had direct experience of the capabilities of the technology, either within education or practice, and subsequently remain unsure as to its potential or its accessibility. This has left us poorly qualified to judge the relevance of CAM to our personal activity; but as the pace of technological development continues to accelerate, how much of this technology will we be able to exploit as designers in the future?

This project attempts to facilitate the activities of designers intrigued by the challenge of this question; however, the mechanism by which this might be achieved has proved more complex and elusive than a mere description of the technology itself. Despite this, the intention behind this book remains an explicit one: to encourage designers from all disciplines to exploit, critically and speculatively, the possibilities of current computer-aided manufacturing within their own practices.

It is beyond doubt that an accessible description of the technology and capabilities of CAM is a necessity to remedy many of the misconceptions currently surrounding it. One element of this book must, therefore, be a descriptive review of the technology itself, in a manner that requires little or no prior experience to comprehend. However, our present relationship with the many processes that shape our environment has become a somewhat distant one, and consequently I am sure that such a description alone would be an incomplete and ineffective means of technology transfer. It would certainly risk ignoring a range of related contexts and issues that directly affect the understanding and application of the technology itself.

This belief has suggested the further content and format of this book: a combination of reference, research and a more dialectic companion, which together seek to place the fascinating and promising technology of CAM within the context of established design and production practice.

Part I
The Pursuit of Novelty

1 CAM: an overview of terms and developments

Computer-aided manufacture has become a rather general and loosely defined term that, in common usage, describes the interfacing of a computer with a machine tool or process to achieve automated operation. Despite the depth of current interest by many designers, CAM is by no means a recent phenomenon, having been initially developed in the 1950s with the support of the United States military. This early form of computer-controlled automation was known as *numerical control* (NC), and allowed the control of modified, but otherwise traditional, machine tools for metalworking applications. New programming languages were also developed specifically for the task, formatting instructions and data into the necessary predetermined sequence. The later derivatives of this system, now commonly known as *computer numerical control* (CNC), still form the foundation of the majority of CAM applications having served industry continually since its inception. Today CNC is used in a much wider range of manufacturing processes, including those involving wood, plastic and ceramic, and the system is operated at a variety of different scales, ranging from the control of a single machine tool to the realisation of complete manufacturing systems.

From the outset, manufacturing industry has sought to expand the use of NC machines to create production methods that could overcome many of the limitations of mechanised mass production and the inherent standardisation of its goods. Considerable research was undertaken to develop flexible manufacturing facilities capable of producing a much wider range of components using a common group of machinery. This was realised in the late 1960s with the advent of the *flexible manufacturing systems* (FMS), which incorporate a number of networked CNC machine tools with robotic handling plant to create a

1.1
A complex mould produced on a desktop machining centre. Image and design: Nat Chard.

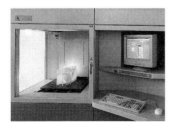

1.2
Solid freeform fabrication: a stereolithography machine and completed prototype. Credit: 3D Systems Europe Ltd.

highly automated manufacturing environment. Unlike the mechanised assembly lines associated with mass production, FMSs allow the production of a diverse range of components with efficiency comparable to mass-production techniques, even for small and medium batch sizes. The further evolution of the FMS has seen manufacturers expand still further the scope of information processing and machine control throughout the manufacturing process, developments which are often categorised under the broad heading: *computer-integrated manufacturing* (CIM).

However, opportunities for designers outside of specialist manufacturing domains to explore CAM were only to be found somewhat later. The development of the microprocessor in the 1970s, and the Personal Computer throughout the 1980s, provided substantial improvements in the on-board processing power and program storage capabilities of CNC machine tools. When these advantages were combined with accompanying reductions in cost, their acquisition came within the reach of small and medium-sized manufacturing enterprises for the first time; often supplementing, rather than replacing, more traditional practices and technology.

The further proliferation of 'desktop' computing was to have a compound effect on the further dissemination of CAM into non-specialist manufacturing environments. The use of computer-aided design and drafting techniques has brought about a universal change in the manner that production information is created, and subsequently CAM techniques have become even more accessible and applicable when coupled with these practices. Previous to this, the use of many CNC machine tools still required specialist programming knowledge that was often non-standardised in its conventions but, as CAD use proliferated, system developers devised further software that could generate programs from computer models alone, enabling the direct use of CAD/CAM techniques by designers and production engineers alike. This freedom has also initiated the development of 'desktop' machine tools conceived specifically for the studio or office. Many of these are often based on conventional machine types, whilst others have been developed from first principles in an attempt to forge the most direct link possible between visual representation and automated realisation. Significant amongst these are *solid freeform fabrication* or *rapid prototyping* processes, which fabricate objects layer by layer in an additive manner under the control of data from a virtual prototype.

The use of CAM has altered the organisation of manufacturing industry significantly in the last four decades, affecting the individual methods and practice of manufacturing as well as creating favourable conditions for inter-disciplinary research and practice. Both current and emerging CAM technology are fundamentally transforming the path from design to production, as the emergence of successful small and medium-size enterprises in manufacturing offer improved access to new forms of manufacturing technology.

Mechanised mass-production methods make extensive use of special tools, moulds and dies specific to individual products, whose necessity remains a significant factor in the large capital investment required to mass-produce any item. Arguably the most beneficial result of CAM has been the revision of this economic relationship, by providing either an efficient means to manufacture pre-production tooling or to eliminate the need for it altogether. Consequently, the lessening of these constraints by CAM techniques provides the automated manufacture of complex components with a diminished reliance on mechanical plant and tooling. This flexible and intelligent form of automation promises to provide designers with a more economic and expressive means to manufacture unique products, whose feasibility is less dependent upon their complexity or quantity. These changing conditions compel designers from all disciplines to re-assess the validity of many manufacturing processes to their domain, as previous assumptions of relevance are unlikely to maintain their validity for long.

'Making a start'

My first experience of computer-controlled machine tools remains an unforgettable demonstration of automation's potential. I was very familiar with the object to be made, a cylindrical metallic container, which was to form the basis of a family of domestic objects. In honesty this brief was merely the initiator for a broader experiment. I had developed a keen fascination with the many roles of the physical prototype as a design tool and, having developed a range of skills in making to explore this, I was curious to see how specialist knowledge of machine tools might further augment this approach. The scene was set for

a b

c d

1.3a, b, c, d

many hours at the controls of a traditional metalworking lathe, as gradually the first prototype took shape.

At first, I found its platonic form quite pleasing, but I soon became acutely aware of how the constraints of both the machine and my skill were forming the largest influence on the design. I intended to make a batch of fifty or so for further evaluation, but it was already clear that far more economical means would have to be used for their realisation. Like many designers I already had a vague notion of the existence of computer-controlled machine tools, but I had always assumed they were the reserve of large-scale manufacturing operations and would be ill suited to my more speculative endeavours. Curious about the truth of this widely held assumption, I decided to visit a firm of precision engineers who specialised in the manufacture of gears, and who also operated a small quantity of computer-controlled machinery.

The reality of this enterprise proved to be one with which I was strangely familiar. Rather than the pristine environment I had imagined, it was a workshop like many I had known, whose organisation appeared to be as much a product of idiosyncrasy as expertise. At the rear of the workshop, beneath a makeshift roof repair of polythene and gaffer tape, the workshop's latest acquisition, a CNC lathe, had been installed without ceremony alongside its more traditional counterparts.

Whilst working at the conventional lathe my final prototype had taken me almost half a day to make, even though the majority of design decisions had been made in previous attempts. Yet, in a display as frenzied as it was precise, this CNC machine condensed my efforts into a two minutes forty-five seconds call to arms (Fig. 1.3).

The work of Sixteen*(makers)

For several years I had, as part of Sixteen*(makers), explored how the act of making could inform the basis of a model for architectural practice. Together with Bob Sheil and Phil Ayres, we sought to define a practice of design neither through the professional demarcations we had trained within, nor through the set of manual techniques our 'industry' displayed, but rather through a revealing of the tacit resource upon which it ultimately relies. The need for such an experiment was founded on our collective belief that neither continuity nor stability could be found within current practices, despite widespread assumptions to the contrary by those who sheltered within it.

1.4

We felt the signs of breakdown were all around us. Even the technical precedents upon which the building industry relied appeared to have been transcended by techniques and forms of organisation in which many designers appeared to have a lessening participatory role. By means of research, Sixteen*(makers) attempted to short-circuit the barriers to innovation and creativity which were generated by this isolation, initially through a series of prefabricated yet also site specific installations (Fig. 1.4).

However, two minutes forty-five seconds of automation had posed a further challenge. With the ability to bring objects into being seemingly at the press of a button, where was the value of the tacit resource that had been invested in creating my prototype? Indeed, could the presence of my prior experience even be considered a form of impediment, obscuring the new realities and possibilities this technology clearly suggested? It seemed significant also that this lesson had not relied on the mechanised machinery and organisation of mass production, but instead was within an enterprise both familiar and accessible – an enterprise rather like our own?

In debating these concerns it was not long before I was reminded of a conversation with Steven Groák, late Director of Research and Development at Ove Arup, who had visited Sixteen*(makers) to view an installation under construction. Once in the workshop it would not be long before he phrased a simple question to disguise a complex dilemma. Noting the reliance our methods placed on the need to design and construct the work ourselves, he asked 'How will you instruct someone else to make it?' We shared dissatisfaction with the inability of many designers to enlist or understand many of the techniques seemingly at their disposal, but in this case his comments highlighted a present and future isolation inherent to our method also. This seemed to implicate us in the perpetuation of the models we sought to transcend.

As such it was an important question to ask, not just for the further development of our own work, but also for any designer who seeks to enlist beyond their personal resource of knowledge and experience. How can the extraordinary capabilities of modern industry, in its many forms, be more fully exploited if we remain so unaware and removed from its reality? Is there a mechanism through which this parallel world of evolving technique would engage with the task of design? At the centre of this debate lies an issue of perennial concern: how can designers continue to be a part of innovation in a time of rapid technical, social and economic change?

I felt these questions, and many others, had been posed again in front of the machine, but this time in a manner that suggested we might be in the midst of a response, if not a definitive answer.[1]

[1]For a description of the work of Sixteen*(makers) at this time see Groák, S. (1996). Board Games. In *Architecture and Design: Integrating Architecture*. London: Academy Editions, pp. 48–51.

2 Division

Computation and mechanisation: the work of Charles Babbage 1791–1871

Charles Babbage belongs to a category of individual rarely encountered within our contemporary condition: the polymath. His range of interests and expertise reflects not solely his evident genius, but rather the nature of the times he sought to understand and influence. He has readily been described as philosopher, mathematician, economist, socialite, writer or inventor, and his thoughts are characterised by a keen search for the connections between each. Today, his achievements, and heroic failures, are most often recounted through his attempts to construct a series of 'Calculating Engines'. These ambitious and visionary projects are well known in forming the basis for his credit as an inspirational figure in the development of modern computing, but their conception and attempted realisation also led Babbage to consider emerging industrialisation in considerable detail, uncovering the rapidly evolving distinctions between designers, makers and manufacturers for which the era has become typified.

Tables

Although Lucasian professor of mathematics at Cambridge, Babbage was also a man of practical concerns that he met with equal measures of intuitive insight and scientific analysis. The problems to which Babbage sought solutions were often everyday ones, and aimed to propagate further the process of industrialisation that he fervently supported. This phenomenon was already well advanced within Britain at the beginning of the nineteenth century; steam power and mechanisation were by then in widespread use, transforming traditional, often rural, artisan crafts into large-scale organised activity within expanding urban centres.

Machine for reproducing sculpture invented by Benjamin Cheverton, 1826. Credit: Science and Society Picture Library.

The task that was to occupy Babbage throughout his lifetime was a direct reflection of these changing conditions, residing not solely within the realms of science but rather in the increasing application of its analytical tools to engineering and industrial practice.

At that time, those engaged in navigation, astronomy and other numerate activities all relied on printed mathematical tables to assist them in the task of 'computation', then a manual process undertaken with dubious levels of reliability. It was also a similarly fallacious process that was used to produce the tables themselves, and Babbage was amongst many in these fields who believed the failings of the tables to be a significant barrier to continuing industrialised progress. This fallibility became an anathema to the developing imperial nation, particularly as belief became widespread that these printed anomalies could be made manifest in the guise of shipwrecks, machine malfunctions or financial irregularities within commerce. Concern about the tables' crisis became so great that the search for a solution became an issue of worldwide interest.

Babbage proposed mechanising the production of these tables, having first identified several opportunities for the introduction of error during their production. The first of these were computational errors in the arithmetic, brought about primarily by the fallibility of human computers, but these were further compounded by additional errors during the transcription of the work for the printer, and still further within the typesetting process itself.

Mechanical calculators had already been produced before Babbage defined his task in 1821. Pascal and Leibniz had both developed working examples in the seventeenth century, but their application and scope was limited to that of a calculating aid for relatively simple arithmetic. Babbage proposed a machine which could not only perform more complex arithmetic, but one which would also emboss the results on a soft-metal block that could subsequently be used for printing. In this manner he believed it would be possible to eliminate all sources of error from the production of tables, realised through the mechanisation of both computation and manufacture. The machines he designed were described as 'Difference Engines', as their operation exploited the established mathematical concept of polynomials, or 'method of differences', to perform complex arithmetic using addition alone.

2.1
One of Babbage's experimental mechanisms. Credit: Science and Society Picture Library.

Enlisting industry

The complexity of the designs was staggering when compared to the majority of mechanical products of the time. The engines were to consist of large numbers of identical parts of many differing types, including gear wheels, cams and levers. Although difficult to quantify from the drawings alone, high levels of precision were certainly required of each individual mechanism, and their assembly into a fully working engine posed a challenge without precedent. Babbage was curious whether the capabilities of mechanical engineering could match these exacting demands, and in 1823 he embarked upon a tour of industrialised centres in Britain and Scotland to acquaint himself better with current practice. The research from these, and other subsequent visits to Europe, formed a comprehensive survey of industrialised activity and organisation, which he published in 1832 under the title, *On the Economy of Machinery and Manufactures*.

His observations are not limited to the practices of any single industry, but instead are categorised through a detailed analysis of emerging practices and organisational forms, the techniques and actions of which are recorded as evidence of a social, scientific and technological transition. Babbage subscribes wholeheartedly to the division of labour, developing further the ideas of Adam Smith in advocating it a necessary model for industrialised and social progress. He speaks for both master and worker, elucidating the benefits of organised production for both, and his research illustrates both existing practices, some of which still exhibit medieval characteristics, as well as those of emerging mechanisation.

Babbage's passion for field-research provided him not only with the statistical evidence of divisions efficiency (the methods of which he is also a pioneer), but also proved the mechanism by which engineers, industrialists and economists alike could readily engage with his ideas. Babbage's popularity and influence quickly rose as a result, particularly as the work was widely read amongst the emerging middle classes, whose capacity to create wealth was invariably a product of innovative technological endeavours.

Amongst its pages we learn of the labours and products of the pin maker and doll maker, the spinner and butcher, the metal founder and printer. Manual, mechanical and natural processes are all identified and accounted for in the search for an analytical proof of the economic benefits of specialisation. Even amongst the flesh and bone of a slaughtered horse Babbage has neither difficulty nor hesitation in revealing the elegance of a commodity fully exploited:

> Even the maggots, which are produced in great numbers in the refuse, are not lost. Small pieces of the horse flesh are piled up, about half a foot high; and being covered slightly with straw to protect them from the sun, soon allure the flies, which deposit their eggs in them. In a few days the putrid flesh is converted into a living mass of maggots. These are sold by measure; some are used for bait in fishing, but the greater part as food for fowls, and especially for pheasants. One horse yields maggots which sell for about 1s. 5d.[1]

Despite his clear enthusiasm for emerging mechanisation, Babbage does not regard the manual skill on which production had been traditionally dependant as redundant, but rather a further resource that is liberated by the division of labour into distinct areas of specialist activity.

Whilst in some instances a complex manufacturing process may be broken down into a series of lesser skilled tasks, Babbage places considerable value in using mechanisation to reproduce artefacts whose original production, say as a prototype or pattern, remained the reserve of a highly skilled artisan. This artisan resource was also instrumental in fabricating the machinery of mechanisation itself, even if it was often at the risk of rendering prized manual skills redundant in the same instance. Through these observations Babbage identifies a new relation between the differing activities of the artisan maker and the mechanised manufacturer, one, which he optimistically believes, is able to serve our desires of both economy and delight.

> The principle alluded to is that of COPYING, taken in its most extensive sense. Almost unlimited pains are, in some instances, bestowed on the original, from which a series of copies is to be produced; and the larger the number of these copies, the more care and pains can the manufacturer afford to lavish upon the original. It may thus happen, that the instrument or tool actually producing the work, shall cost five or

---

[1]Babbage, C. (1835). *On Economy of Machinery and Manufactures.* 4th edn. London: Charles Knight, p. 395.

even ten thousand times the price of each individual specimen of its power.

Operations of copying are effected under the following circumstances:

By printing from cavities.	By stamping.
By printing from surface.	By punching.
By casting.	With elongation.
By moulding.	With altered dimensions.[2]

Babbage's account presents an intriguing image of the emergent factory as a realm of often-familiar actors and scenarios, each caught in a process of transformation that alters their scale, speed and ultimate industrial capability. It is an era that begins with each artefact, however trivial, as a unique entity specific to its author, yet rapidly transforms into a mechanised age in which the original is put to service as the tool, the mould or the master pattern: fundamentally the progenitor of its type.

Designer and maker

Despite his vision and enthusiasm Babbage failed in his attempts to construct a fully complete working calculating engine within his lifetime. The reasons for this failure are many and varied; being both a product of limitations within nineteenth-century engineering practice, endless vagaries of funding and, in no small part, the individual relationships between the characters involved. The respective influence of these factors continues to be widely debated.

The construction of 'Difference Engine No. 1' had begun with sincerity and optimism in 1824, when the task was entrusted to Britain's then most renowned precision engineer Joseph Clement. However, it was not long before the flaws within Clement's and Babbage's working relationship were to become evident, as delays and disputes became endemic to the project. Although both had undoubted respect for each other's considerable abilities, the complexity of the undertaking created numerous occasions for dissatisfaction and ultimate mistrust. New forms of machine tool had to be constructed to manufacture the engine's parts,

---

[2]Babbage, C. (1835). *On Economy of Machinery and Manufactures.* 4th edn. London: Charles Knight, p. 69.

themselves significant contributions to manufacturing, and Babbage became concerned that Clement sought to increase his own knowledge, standing and wealth even at the sacrifice of the initial goal. Babbage, for his part, compounded these problems with his intermittent supervision of the project during his European travels and his failure to clearly define funding targets with the government at the project's commencement. Despite the slow rate of progress Clement did, in 1832, finish a portion of the engine, being one-seventh of the final machine. It contains almost 2000 individual parts, and its successful operation was the most powerful evidence of the feasibility of Babbage's design within his lifetime. By now, however, the project was the cause of escalating and increasingly bitter financial disputes between Clement, Babbage and the government to such an extent that relations between all parties became irreconcilable, and consequently no further construction work on the engine was to take place beyond 1834.

By now it was also clear that the undertaking of designing a new form of calculating engine had distracted Babbage's attentions. These even more ambitious machines were not confined to pre-determined calculations, but proposed to allow its user to program a wider variety of functions using punched cards (inspired by Jacquard's automated loom developed at the turn of the nineteenth century). This family of unrealised designs, known as the Analytical Engines, is now universally considered as the forerunner of the modern general-purpose computer, and Babbage's description of its architecture can be recognised within digital computers to this day.

The failure to fully realise Babbage's ideas during his lifetime has ensured a legacy far more inspirational than didactic, providing valuable concepts but rarely the practical means for future developments in computing. With none of the calculating engines fully realised at the time of Babbage's death their capabilities seemed destined to remain one of technology's most intriguing and promising riddles, and one which was to create a moment of discontinuity in both the development of computing and, to a similar extent, automated manufacture.

Even during his work on the Analytical Engines, Babbage still pursued a solution to the table's crisis that had occupied him years earlier. His experience in developing the design of the Analytical Engines had also suggested a revised and simplified form of Difference Engine, which he designed between 1846 and 1848. Its realisation, he believed, would be far more feasible than its earlier counterpart, but persuading the government of these claims was something Babbage undertook with understandable pessimism, yet with an obvious moral obligation to fulfil his earlier contract. By now Babbage's esteem and influence in government circles was clearly spent and his proposal was rejected without the slightest consultation as to its merits. Babbage's disappointment was to envelop him until the end of his life in 1871.

Vindication

In 1991, as an attempt both to celebrate Babbage's achievements, and to contribute to the contemporary debate regarding the feasibility of his ideas, the Science Museum in London used surviving drawings to successfully construct Difference Engine No. 2. It would be no surprise to Babbage to find that its belated fabrication has been a product of both computer-controlled automation and manual skill, together comprising a hybridised resource whose capabilities were previously without precedent.

2.2
Reg Crick, one of the team of engineers that constructed Babbage's Difference Engine No. 2 at the London Science Museum. Credit: Science and Society Picture Library.

The construction of Difference Engine No. 2 is a further testament to the vision of Babbage, but intriguingly is not the indictment upon nineteenth-century mechanical engineering that some had expected. For this retrospective realisation modern computer-aided manufacturing methods were used extensively, particularly for the large numbers of interchangeable components required, but the benefits of these contemporary technologies were exploited more for purposes of economy than accuracy alone, providing access to a new form of efficiency rather than a new level of precision. To allow this comparison, existing examples of Clement's work were measured using modern equipment to determine manufacturing tolerances of the period, and for the re-construction project efforts were made to ensure that the accuracy of any machining did not exceed that exhibited within initial attempts to construct Difference Engine No. 1.

The success of the Science Museum's re-construction certainly lays to rest the notion that Clement's workshop was not capable of producing work to the required standard, although to some extent this had already been suggested by the working fragment of Engine No. 1. The new engine is perhaps greater evidence of the logistical challenge of manufacture, and focuses our attention upon the organisational constraints surrounding production during Babbage's time rather than the technical capabilities of production alone. In his attempt to construct the Calculating Engines Babbage sought artefacts beyond the individual capabilities of either artisan or industrialist alone; too complex to practically realise by manual means, yet too unique to exploit emerging techniques of mechanised reproduction. With their countless interchangeable and repeatable parts, their creation relied on a means of production whose characteristics, at that time, could only have been suggested within the operation of the Calculating Engines themselves, simultaneously compressing time and material in a pre-programmed act, and creating a physical topography from a stored mathematical abstraction.

In 1999 the Science Museum's project completed its final phase, by constructing the printer that creates the physical output of the difference calculations. Contemplating this most sublime of objects, one can almost hear the prophetic voice so long denied by its delayed and untimely birth, and audible now only as an isolated archaeological precedent to many subsequent generations of computer-controlled manufacture.

However sophisticated, and even if realised, the fragmented mechanics of Babbage's world were destined to redundancy by the speed and simultaneity of electricity, and the adaptive and flexible automation it was later to create when enabled by the digital computer and servomotor. Yet his work also stands independent of its influence on computing, displaying unequivocal evidence of a designer's relationship with manufacturing that was more beneficial and influential than the saga of the Calculating Engines suggests alone.

Babbage desired equally to develop the processes that could bridge the gap between skill and mechanisation; most significantly his work recognises the role of the computer program as the authored 'instruction' necessary to translate mechanical production into a more flexible form of automation. For Babbage manufacturing exists almost as a medium; aware of the individual tools, techniques and organisations that make it up, he attempted to work both within and beyond them, combining imagination and constraint to exploit existing capabilities and to develop further ones. His knowledge of industry's capabilities is a necessary pre-cursor to its appropriation for his own extraordinary needs.

Prototypes

Babbage's understanding of the original, as something to replicate and exploit, is very much in keeping with his subscription to the division of labour, and sets the scene for future mass production and consumption. Yet the act of creating the prototype itself, as a creative and speculative act, does not figure so directly in his accounts. Reviewing his polemic of production this seems both predictable and explainable, yet if we peruse his tools, notation and models we witness a curious enigma. Here is evidence of a designer at work, testing through a variety of mediums the validity of his assertions in moments beyond the cold logic of specialisation.

Babbage's personal approach to the task of design seems highly influenced by his involvement in manufacturing, but this insight alone does not explain the variety of knowledge, skill and dexterity exhibited within his own design methods. Despite Babbage's clear affinity with analysis he also appears an avid empiricist. Significantly, his knowledge of metalwork and machining in particular was not derived from his documentary research alone, but also from his own

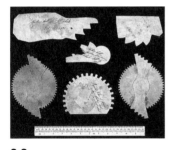

2.3
From abstraction to realisation: Babbage's card models for Difference Engine No. 1. Credit: Science and Society Picture Library.

practical work and experience. Ornamental turning, often carried out on elaborate and decorative lathes, was a fashionable hobby for a gentleman at that time, and Babbage made a variety of working models and mechanisms to formulate his designs. The benefits of this tactile experimentation were further augmented by his system of Mechanical Notation, which used a variety of symbols to describe complex moving mechanisms. This novel form of representation was conceived by Babbage specifically to assist in the design of the Engines, and provided the means to visualise the interactions between connected components that engineering drawing alone could not express so readily.

The uniqueness of the Calculating Engines, each one a prototype, demanded a combination of precision and efficiency which, until recently, was only economically feasible with mass-production techniques that were applicable to the creation of standardised multiples alone.[3] Recent developments in computer-aided manufacture are focused once again on the economic creation of the prototype and its transformation into pre-production tooling for subsequent manufacture. Even the automotive industry, previously the benchmark in standardisation, seeks to achieve the customisation of its products and reduce its reliance on mass production. The economic advantages of this technology are straightforward to state, but the true significance of computer-aided manufacturing lies within its ability to initiate a rare reaction between conception and execution, whose implications go well beyond those which can be explained using economic principles alone.

The desire to realise unique objects and constructs continues to be shared by designers and manufacturers alike. Our desire to customise the environments and products of our world is a human one; so perhaps the techniques that realise this ought not to be classed simply as industrial tools, but rather as an extension of an existing medium of expression. By striking at the previously limiting conditions of standardised mass production, computer-aided manufacture allies the idiosyncrasies and surprises of design tantalisingly close to the realities of production.

---

[3]By the end of the project in 1834 the total cost to the government was $17 470. The size of this may be gauged from the cost of the steam locomotive *John Bull,* built by Robert Stephenson in 1831 for $784. Source: Swade, D. (1991). *Charles Babbage and his Calculating Engines.* London: Science Museum.

As a designer, to act upon this inspiration today remains a problematic task. With the definition of our role diminished by two centuries of division and sub-division of labour, the knowledge left in our possession seems insufficient to re-cast ourselves in a more self-assured role. Is it possible to appropriate from domains of production outside of our present experience, or will this become the mechanism of our own demise, distracting us from the value of our own position as designers?

Existing perceptions of mechanised mass production are in danger of obscuring other emerging partnerships between individuals and industrial resources, which allow technological systems to be explored as a medium of creativity. Paradoxically, there is a sense that Babbage understood this condition despite his avocation of more fragmentary means. Consideration of the potential reconciliation between design and production remains important not simply to go beyond the misconceptions of division, but also to consider and confront the increasing specialisation upon which so many current practises rely. Though it may also be a necessity in part, the unchecked rise of the expert threatens, like mechanisation before it, to further separate conception and execution into disparate and conflicting activities. Too many designers exhibit the symptoms of the machine operator, unable to intervene or influence the process they allegedly control, ironically excluded by the very technology and organisation they seek to exploit.

2.4
The printing and stereotype apparatus for Difference Engine No. 2. Credit: Science and Society Picture Library.

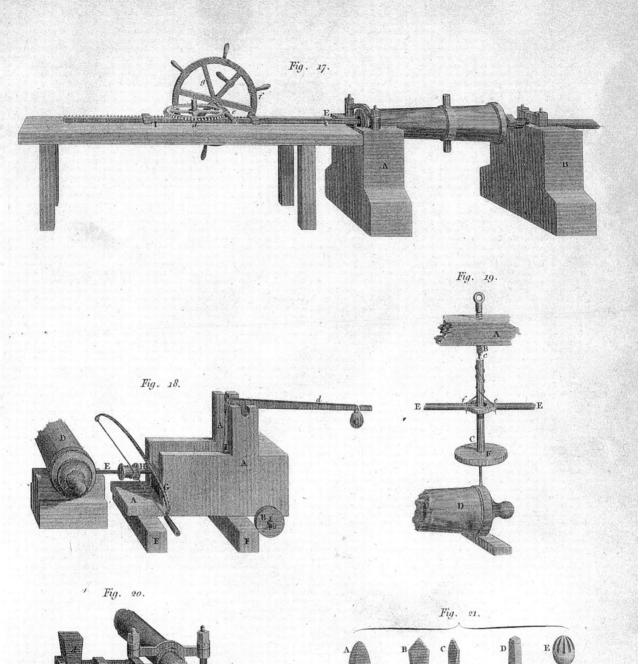

Fig. 17.

Fig. 19.

Fig. 18.

Fig. 20.

Fig. 21.

m. delin.

Published as the Act directs, 1808, by Longman, Hurst, Rees & Orme, Paternoster Row.

Engraved by Wilson

3 The primitives of manufacturing: a select history

3.1
Credit: Science and Society
Picture Library.

Cannon boring, 1808. Plate from
Rees's Encyclopaedia, volume 2.
Credit: Science and Society
Picture Library.

The tacit resource

When Thomas Newcomen (1663–1729) produced his first successful steam engine in 1712, more than a hundred years before Babbage published *On the Economy of Machinery and Manufactures*, it was without the use of any of the machine tools taken for granted today. At that time, Britain faced a timber crisis as the combined demands of the ship industry, a growing iron industry that used charcoal, and domestic fuel use began to deplete existing stocks at an unsustainable rate. The use of coal as fuel and iron as a structural material offered a tantalising solution, but at that time coal extraction was limited as horse-driven pumps could not drain deep mines and no satisfactory way of smelting iron with coke was available. The endeavours of Abraham Darby, who may have begun smelting iron with coked coal at Coalbrookdale, Shropshire, as early as 1709, and Newcomen's first successful pumping engine at Dudley Castle in 1712 were distinct but interrelated achievements, which were to form significant catalysts to the extraordinary industrial growth of the next two centuries.

The cylinders of Newcomen's first engines are believed to be of cast brass, measuring 2500 mm in length and with a bore of approximately 600 mm in diameter. Although simple boring mills to produce cannon had been continually developed since the sixteenth century, these were capable neither of producing work of this size nor to the required accuracy; only laborious and highly skilled hand finishing using whetstones could achieve a satisfactory result.[1] Many of the simple hand tools available to Newcomen and his workers demanded considerable dexterity in their use,

---

[1]Rolt, L.T.C. (1965). *Tools for the Job: A Short History of Machine Tools*. London: Batsford.

requiring actions learnt consciously through an apprenticeship of hand, body and mind yet, once acquired, expressed with intuition and fluency. Perfected over time, mastery of any task would be tested against each technological challenge anew, and the experience gained in its accomplishment would be held, to a large extent, as the privileged memory of the artisans directly involved; unable to be fully expressed through language, drawing or convention alone. As a result, Newcomen's engine was like many products of the time, the result of an empirical approach to a clearly defined problem, the solution to which relied heavily upon craft skills. The success of this approach relies heavily on the transfer of the tacit knowledge contained within these craft traditions, providing the means by which techniques and past experience could be invested into the task by non-verbal means.[2]

Accessing this tacit resource was essential within the empirical approach exhibited by Newcomen and his contemporaries, particularly as engineering drawing did not then exist as a standardised means of communication. Their invention relies on an 'empathetic engagement' of intuitive thought and action, the outcome of which is revealed by the behaviour of materials in an unambiguous reality. This 'trial and error' approach would soon prove inadequate, however, as an emerging engineering profession attempted to apply a growing body of research based on scientific principles and mathematical analysis, a task which often necessitated the collaboration of several different practitioners, each with their own specialist knowledge and experience.

Collaboration

Few were more influential than James Watt (1736–1819) in crossing the divide from an engineering practice derived from medieval models to one equally grounded within knowledge of scientific principles and expressed through emerging conventions. Watt was equally familiar with both forms of practice, as his early experience as a scientific instrument maker at Edinburgh University furnished him with the practical skills and scientific knowledge to research empirically and analytically. It also, by chance, provided him with the occasion to revise Newcomen's early contribution to industrialisation.

---

[2]Hendersen, K. (1995). *The Visual Culture of Engineers from the Cultures of Computing.* (S.L. Star, ed.) Oxford: Blackwell, p. 206.

Whilst repairing a model of a Newcomen engine, which was used for teaching purposes at the University, Watt became aware of inefficiencies in its operation. He duly embarked upon designing an improved form of engine, and was assisted in its development by knowledge of the nature of heat unknown to Newcomen, which was provided in part by the work of his academic colleague, Joseph Black. Despite illustrating the success of his engine experimentally, its reliable production at full-scale proved more difficult, requiring manufacture of a scale and accuracy beyond that found in any machine shop of the time. Five years would pass before Watt, who by then was in partnership with entrepreneur Matthew Boulton, could instruct John Wilkinson to undertake the task, enabling Watt's improved steam engines to power the continuing industrial advance both at home and abroad.

Watt's struggles in realising his design mark the beginning of an era in which an increasing number of products owe their existence to specialist manufacturing techniques outside of the previous domains of the artisan workshop. Watt clearly understood the limits of even his own exceptional manual skills, and realised that future technological advance would be increasingly dependent on partnerships with other innovative modes of emerging production. His development of the engineering drawing was a significant step towards such forms of collaboration, providing an unambiguous means to record not only explicit instructions but also tacit insight. Watt's pioneering use of engineering drawings enabled their use as tools for design, development and production, and his work became characterised by the routine use of many specialist firms to realise any single project.

The mechanical resource

In Britain, the first half of the nineteenth century saw an accelerated development of many machine tools and their products. Engineers including Clement, Fox, Maudsley, Nasmyth and Whitworth all made significant contributions to the development of machine tools in Britain. Many traditional machining and forming processes such as cutting, bending, drilling and forging were enhanced by steam power, enabling work of a speed and scale previously unimaginable by manual means. Early machine types, such as the lathe, were the subjects of constant innovation offering significant improvements in precision and versatility, while new types,

3.2
Precision gauges by Whitworth.
Credit: Science and Society
Picture Library.

3.3
Screw rolling machine, 1851.
Credit: Science and Society
Picture Library.

such as milling machines and gear cutters, were used successfully for the first time. In addition, existing techniques and practices were regulated and codified by a growing engineering profession, notably the standardisation of screw threads and the description of gear tooth geometry. All of these developments combined to ensure a steady increase in the accuracy and capability of machine tools, greatly increasing the range of industrial and domestic products available. At the end of this period many of the machines that still shape our products today were in widespread use, albeit in simplified forms. These 'primitives' formed a basic palette of material shaping and forming techniques, which although extended through time remains discernible in industry today – even amidst the vast range of current computer-aided manufacturing processes.

These developments alone, however, prove to be only a single facet of emerging mechanisation. Although the steam powered workshops such as those of Watt or Joseph Whitworth exhibited the highest levels of innovation and technology, they did not yet display the organisation by primarily economic principles that we have later come to associate with twentieth-century mass production. Workshops of this period fabricated much of the machinery required by early factories (the textile mills of England and America are some of the earliest examples), but their own production strategies still bore a strong resemblance to more traditional artisan methods.

Despite the division of labour, and the increasingly complex machinery at industry's disposal, the production of many engineering products was still highly dependent on manual skill. This remained a necessity both for the many manufacturing operations that had not yet been mechanised, and often for the operation of the machines themselves, which although self-regulating to some extent, did not completely subjugate skill in their use. Within many engineering workshops the tacit resource of the artisan workforce created a flexibility and diversity to the firm's capabilities that automated mechanisation alone could not exhibit. However, despite the advantages of this condition, developing capitalism would consider this lingering medieval trait a clear constraint to its goals, and the lessening of this human dependence became a priority for many entrepreneurs. The desire to develop a production strategy less dependent on skill was shared by industrialists in Britain and Europe, but its earliest resolution would appear in a nation whose demands were perhaps more pressing than any other at that time: America.

'The American system'

Industrial development in eighteenth-century America had been hindered by a British government keen to exploit the colony solely as an export market. Even after American imports of iron were permitted in 1750, an embargo on the colonial manufacture of steel still remained, and after the Declaration of Independence in 1783 the British government banned the export of any British machine tools or the emigration of any individual associated with the iron or manufacturing trades. However, these punitive measures were to prove no obstacle to the industrial advance of an emerging nation rich in physical and human resources. Emigrants from both Britain and Europe played their part in an initial period of technology transfer that soon blossomed into home-grown innovation.[3]

Eli Whitney (1765–1825) typified this time. An American by birth, though drawing on ideas and expertise from world-wide sources,[4] he is known not so much for his inventions, but for his desire to develop manufacturing into an organised system of production. Whitney's 'American System' was to play a significant part in transforming the United States into an industrialised nation, and in so doing defined the nature of manufacturing industry into the next century.

Whitney's first contribution was to the textile industry at the end of the eighteenth century. His 'cotton gin' of 1793 enabled use of the short staple cotton plant whose seeds had previously been uneconomic to remove from the crop, and it quickly enabled the United States to become the world's most productive textile industry. Patent wrangles brought Whitney little profit, however, and in 1798 his entrepreneurial eye was caught by another national concern: increasing problems in arms production.

Firearms of the period, whilst unreliable at best, suffered the additional weakness that their individual parts were not interchangeable. Each component of the rifle was fitted by

---

[3]Rolt, L.T.C. (1965). *Tools for the Job: A Short History of Machine Tools.* London: Batsford.
[4]Bassala, G. (1988). *The Evolution of Technology.* Cambridge: Cambridge University Press, p. 32.

hand to form a unique assembly and, despite any standardisation in overall design, similar parts from one rifle would be useless to another. For a pioneering population in an often-hostile landscape, this created an obvious compromise to personal and national security.

Whitney's proposed remedy to this was a system of 'interchangeable manufacture', in which rifle parts would be manufactured with sufficient precision to allow their assembly in any combination. This would also facilitate later repairs using parts that were similarly standardised in their design and dimensions. The further appeal of Whitney's system was the intention to reduce the unit cost of the rifles to below that of existing artisan production methods. The benefits were clear to both government and industry. Realising the potential economic and political rewards, Whitney approached the task with staggering confidence, motivated as much by impending bankruptcy as by inventive zeal, and successfully securing a contract from the US government for 10 000 rifles before having even constructed a facility to manufacture them.

Unfortunately, Whitney's enthusiasm did not match his initial rate of production, and it was to take Whitney eight years to deliver the full quota, six years longer than the original contract had specified. Having failed to meet the first year deadline Whitney pacified the government, then on the brink of civil war, with announcements of his 'new system of manufacturing'. Whitney re-stated his intention to utilise mechanised production methods to create artefacts whose component parts could be readily interchanged. From the outset this technical and organisational challenge promised considerable financial reward to all sectors of manufacturing, as it suggested a form of mechanisation that transcended the need for high levels of artisan skill, and effectively allowed the complete re-organisation of even complex precision manufacturing processes under predominantly economic principles.

Such was the continued appeal of the concept to government that the realities of Whitney's poor past performance were ignored. Whitney not only continued to receive generous financial backing, but also became an influential government advisor on a range of social and economic issues. The strength of his personality, and his willingness to exploit the media, saw Whitney transformed from man to myth within his own lifetime. Recent research has cast doubts on the true success of Whitney's first attempts at

3.4
Broaching the surfaces of steel rifle parts (twentieth century). Using a multi-toothed cutter, parts are machined in a single pass. Reproduced from W. Wilson Burden (1944). *Broaches and Broaching*. New York: Broaching Tool Institute.

3.5
An automotive part (foreground)
and the cutting tool used to
machine it. Reproduced from W.
Wilson Burden (1944). *Broaches
and Broaching*. New York:
Broaching Tool Institute.

'interchangeable production', but it remains the case that his public championing of the technique, and his adoption of organised management, production and quality control strategies, were all significant steps towards that goal.[5]

Transfer

The American firearms industry, which consisted of both government-owned and private contractors, continued to develop new mechanised techniques to improve precision. Manual operations such as filing, for example, were substituted for new machining processes, including milling and broaching. Additional types of drilling and profiling machines were developed, as were automatic stops to control dimensions, and shaped cutters to reduce the number of machining operations during manufacture. Armouries from New England and Virginia operated a policy of continual and shared development, facilitating technology transfer and accelerating development. This activity became focused on the national armoury at Springfield whose governmental status allowed its staff access to other workshops.

From the 1840s on, the techniques of interchangeable manufacture were quickly disseminated to other manufacturing centres in the US. In many instances, this process of diffusion was often enabled by workers initially trained in the firearms industry lured by the greater rewards of new ventures. America also began to manufacture the specialist machine tools that had been developed experimentally over the same period, allowing their use in the production of a much wider range of goods. American machine tool manufacturers displayed their wares for the first time in Europe at the Great Exhibition of 1851, fittingly in Paxton's marvel of prefabrication, the Crystal Palace. Such was their success at the exhibition that the British government sent a commission, chaired by Joseph Whitworth, to tour America and report on the considerable advances that had been made there. As a consequence, when the British government re-equipped its arsenal at Enfield, it was to make extensive use of American machine tools.

The development of armaments remains a situation where innovation is pursued at costs far greater than the conventional wisdom of profit and loss dictate, and it is clear that

[5]Smith, M.R. (1981). *Eli Whitney and the American System of Manufacturing from Technology in America (1986)*. (C. Pursell, ed.) Cambridge: MIT Press, reprint.

3.6
Royal Small Arms Factory,
Enfield, London, 1861. From the
Illustrated London News. Credit:
Science and Society Picture
Library.

the implications of Whitney's work to national security accelerated the realisation of the system. That it became so widely disseminated, however, illustrates not simply its ability to increase productivity alone but its ability to address wider production problems of the period, not least of which was obtaining labour. Whilst the US did not suffer from a labour shortage, the range of opportunities available to the population created a migratory workforce, easily lured to new and better-remunerated ventures. Mechanisation's ability to 'de-skill' operations that had traditionally been the domain of craftspeople of many years training enabled the exploitation of a wider range of the population, facilitated lower rates of pay, and was more tolerant of high workforce turnover.

The growth of mechanisation: a global strategy

'Interchangeable Manufacture', or the 'American System', continued to influence the emerging manufacturing industry that would soon command a global marketplace. Its evolution into the assembly lines epitomised by Henry Ford's Baton Rouge automobile plant was advanced as the twentieth century dawned, enabling the commercial production of a vast range of new products consumed by a rapidly expanding and empowered population. Designers from all disciplines were also faced with new opportunities, as the technical and social implications of mass production created unfamiliar challenges for those who wished to enlist it. The hardware of manufacturing was becoming

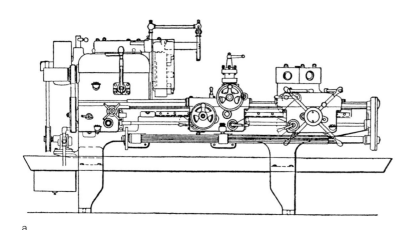

a

3.7a, b, c, d, e
Time and Motion. These
schematics describe the use of
a manual turret lathe to machine
a brass cover. Note the
allowance made for 'tool setting,
changing and fatigue'.
Reproduced from *General
Engineering Workshop Practice*.
This extract by H.C. Town.
London: Odhams, reprint, 1959.

b

d

FIRST OPERATION				
OPERATION	SPINDLE SPEED. R.P.M.	SURFACE SPEED. F.P.M.	FEEDS PER IN.	TIME IN MIN.
Chuck				·20
1 Bring up 1 . . .				·10
Rough turn and bore .	250	200	80	·36
Bring up 2 . . .				·10
2 Finish turn and bore .	250	200	80	·36
Bring up 3 . . .				·20
3 Rough face . . .	250		Hand	·32
Bring up 4 . . .				·20
4 Finish face . . .	250		Hand	·32
Remove work, etc. . .				·25
				2·41
For tool setting, changing and fatigue, 25% . . .				·6
Total (approx.)				3 min.

c

SECOND OPERATION				
OPERATION	SPINDLE SPEED. R.P.M.	SURFACE SPEED. F.P.M.	FEEDS PER IN.	TIME IN MIN.
Chuck				·20
Bring up 1 . .				·10
1 Rough turn . . .	250	200	80	·36
Bring up 2 . .				·10
2 Finish turn . .	250	200	80	·36
Bring up 3 . .				·20
3 Rough face . . .	250		Hand	·32
Bring up 4 . .				·10
4 Finish face . .	250		Hand	·32
Bring up 5 . .				·10
5 Form neck, bevel and rad.	250		Hand	·10
Change speed, bring 6 .				·20
6 Thread	150	40	Hand	·10
Remove work . .				·20
				2·76
For tool setting, changing and fatigue, 25% . . .				·69
Total				3·45

e

increasingly specialist, and as a consequence, designing for mass production began to require knowledge of processes and materials that were unfamiliar to designers from non-engineering backgrounds. New mechanised techniques and developments in man-made materials, including polymers and metal alloys, would mean inevitable isolation for designers who chose to remain conversant solely in the language of traditional craft.

The design of many everyday products became founded on a prior assumption of their impending mass consumption, particularly as their manufacture became reliant on vast capital sums to create and maintain the mechanised equipment that was often unique to that product alone. Creating items by these means contrasted greatly with the almost spontaneous manner in which the artisan could commence production with only the simplest sketch or conversation as preparation, and designing for industry effectively became a discipline in its own right. The emergence of the industrial designer was not simply a product of mechanised developments alone, but had gathered pace since the earliest separation of design and production brought about by the division of labour. (The production of eighteenth-century pottery by Josiah Wedgwood is one of the earliest examples.)[6]

Despite concerns of the artisan, the proliferation of mass-produced items soon provided ample evidence of the innovative and expressive designs that its methods alone could realise. Many items, such as the automobile, were becoming products not born of technology alone however, but of an associated cultural strategy that ensured their promotion was not limited to claims of functional or aesthetic superiority alone, but was also inseparable from images of enviable status and lifestyle.

Designers outside of manufacturing faced a difficult dilemma. Standardisation had become a necessary prerequisite for any collaboration with manufacturing industry, as designs became a direct expression of the mechanised methods used in their production. Those who wished to embrace this phenomenon found the validity of its techniques questionable when applied to smaller volumes of production, whilst those who rejected mechanisation would find that artisan skill was becoming a rare and

---

[6]Forty, A (1986). *Objects of Desire*. London: Thames and Hudson, p. 29.

3.8
Credit: Science and Society Picture Library.

increasingly expensive commodity, whose application was becoming a privilege.

Following Marx's predictions, mass production became the reserve of increasingly large firms and global corporations, and much of the making of everyday items – new modern necessities – became a virtually invisible activity, revealed only through iconic or clichéd images within media and education, and rarely through the personal experience of the individual.

Outstanding value for money

PREFECT

4 Towards an age of mass customisation

4.1
Food production tooling: a pasta die. Credit: Science and Society Picture Library.

4.2
A commercial mould for Cadbury's 'Giant Bar'. Credit: Science and Society Picture Library.

I think that cars today are almost the exact equivalent of the great Gothic cathedrals: I mean the supreme creation of an era, conceived with passion by unknown artists, and consumed in image if not in usage by a whole population which appropriates them as a purely magical object.

Roland Barthes 'Mythologies'[1]

For Barthes 'The New Citroen' is a sublime mystery and the icon of its age. The tangibility of its ownership and the complexity of its tectonic belie the hectic domains of assembly and fusion that brought about its realisation. Despite its 'absence of origin', we know this is a product of our society today, of a complex partnership of knowledge, organisation and desire that has founded this very particular technology. As consumers, the fruits of these unseen labours have become ours to subscribe to, but the sophistication of its technology appears to have created a distance from our own experience of how things come into being. Society's products are created in a number of increasingly specialist production environments using a vast range of unrecognisable tools and techniques; even allegedly 'simple' products, such as the common chocolate bar, are produced in unique and complex facilities that seem far removed from the humble alchemy of our own kitchens.

An elusive partner

The emergence of the manufactured product during the nineteenth century caused designers from many disciplines to question the validity of established artisan methodology

---

[1]Barthes, R. (1957). *Mythologies*. Translation Annettte Lavers (1993). London: Vintage, p. 88.

in addressing the changing demands of society, and opinions were often as divided as they were extreme. Both William Morris and John Ruskin saw the mechanisation of production as a threat to all creative endeavours; extolling the virtues of handcraft, and adopting a model of production more akin to the medieval craft guilds, Morris created a social-aesthetic theory in direct opposition to industrialisation. Yet despite the promotion of similar thought through a popular Arts and Craft movement other, more collaborative, attitudes towards the capabilities of industry were to be more influential in shaping the tectonics of the built environment and the objects within it.

One of the characteristic features of much of the last century's design has been its technological polemic; one that has alluded to, though never attained, a symbiosis with manufacturing and industrialised techniques. The possibilities of this union formed a foundation on which many of the aims of the Modern movement are firmly set. Although influencing a range of design disciplines, this is particularly evident in the discourses of architects Walter Gropius and Le Corbusier. Like Morris, Gropius believed in a 'common citizenship of all forms of creative work'[2] but in this instance his rhetoric embodied an imperative to ensure that architects and designers of the period worked in close collaboration with industry. In founding the Bauhaus, Gropius created an environment in which all its students were expected to become familiar with factory methods of production. Paradoxically, this would be achieved not through the rejection of handcraft skills but, on the contrary, through their acquisition; these traditions were considered a practical and relevant mediator in the understanding of the relationship between conception and production. For Gropius, standardisation and rationalisation were not new obstacles to creativity, but were a traditional and necessary structure by which the future products of the machine would gain meaning and artistic expression.

Apprenticeship was not the only means through which union was sought. Le Corbusier was also inspired by the products of industry, most famously the automobile and the aeroplane, yet his approach acknowledges a need for architecture to seek its own forms and techniques outside of the domains of manufacturing. Both practitioners sought to enlist new technology to the cause of architecture

---

[2]Gropius, W. (1935). *New Architecture and the Bauhaus.* Translation P. Morton Shand. London: Faber and Faber.

through the exemplar of the machine and the mechanisation of production, demanding an expansive revision of building activity enabled by emerging industrialised technique. Gevork Hartoonian, noting the implications of these positions, describes the embodiment of this search:

> Technical positivity manipulated the figurative aspects of architecture. The result was reflected in the idea of object type, a model or prototype whose repeatability is a dimension of mass production and whose peculiarity is a function of building types.[3]

Modernism continued to be influenced by these desires into the 1950s where they were allied with the mammoth task of re-building and expanding the post-war landscape of Europe and America. This proved a harsh critique, whose requirements highlighted a problematic partnership between the ideals of modernism and the realities of emerging industrialised building techniques. Despite the enthusiasm for the qualities of the manufactured product, little of the actual technology of mechanised manufacture could be applied directly to the construction site; instead the building industry had adopted practices more influenced by theories of scientific management to obtain greater organisation of labour within the building process itself. Many developments in new building technology, including the systemisation and prefabrication of its products, were grounded in contractor-based motives to meet exceptional demand of the period.

As a result much of the post-war landscape became illustrative of a transition from an early rhetoric that acknowledged tacit practice within the building process to those that effectively replaced it with a desire to state explicit rules.[4] The early tenets of Gropius and Le Corbusier had relied upon a metaphor of the machine that masked a failure to define which dimensions of manufacturing industry could practically be allied to a new architecture, or indeed which organisational forms might be required to realise this. The failure of much post-war industrialised building illustrates the struggle many practitioners faced in defining those elusive aspects, and also the inability to fully comprehend the nature of their own industry in a time

---

[3]Hartoonian, G. (1999). *Ontology of Construction.* Cambridge: Cambridge University Press, p. 41.
[4]Groák, S. (1983). Building Processes and Technological Choice. *Habitat International*, Vol. 7 No. 5/6, p. 363.

of rapid change. The debate that post-war industrialised building initiated has had widespread implications since that time, tempering much of the previous technological optimism and subsequently forming critical attitudes within the building industry towards the possible value of technology transfer in the future. Whilst many designers are routinely inspired by the opportunities emerging science and technology has suggested, the rapid influence of these ideas within theory and research is often in stark contrast to the methods available to many contractors.[5]

Goals

Many of the conclusions reached through the experiments of the post-war era are sensible and considered, not least of which is the understanding that manufacturing industry and the building industry are radically different, both in their goals and the means by which they are achieved.[6]

Mass production has become characterised by the production of large numbers of standardised products exhibiting minimum variation. Its function is grounded in a separation of design and production, which is manifest in the technology and architecture of the mechanised assembly line. Central to its success is a desire to reduce the necessity for manual skill, and its associated uncertainties, within the manufacturing process. Mass production enterprises of the mid-twentieth century display an awe-inspiring commitment to the hardware of production, harnessing the labour of a massive local workforce and constructing a complex supply chain of raw materials and auxiliary services. These enterprises, focused around the fixed geography of 'the factory', have shaped and defined the social and economic fabric of entire regions for prolonged periods. Mass production has sought consistency in the behaviour of its products, its workforce and its consumers. Only by attempting this has the predictable and reliable exploitation of a global marketplace become a reality.

In vivid contrast, the building industry appears to operate in a distinctly different and rather implausible manner. Each construction project displays a level of uniqueness quite

4.3
Credit: Science and Society
Picture Library.

---

[5]A resolution of this condition was attempted through a collaboration between Sixteen*(makers) and Neil Spiller by creating the 'Hot Desk'. See AD issue Architects in Cyberspace. London: Academy Editions.
[6]For a further comparison of the two industries see Groák, S. (1992). *The Idea of Building.* London: E. & F.N. Spon, p. 137.

unlike the products of mass production. Nomadic and flexible, it utilises a wide range of technological solutions to realise any particular building type, whose characteristics are dependent upon many conflicting factors, including the physical nature of the site, the skills of the available workforce, prevailing ideologies (both of design and consumption), knowledge of materials and technical precedent. Fragmentation and uncertainty are endemic to the industry.[7] Each project demands complex communication between a unique constellation of participants, each with their own vested, and often seemingly incompatible, interests. The building itself, regardless of whether traditional or emerging techniques are used in its realisation, will represent just one of an infinite number of permutations of connection and enclosure and, whilst its desired performance will have been specified in advance, its performance in use is neither completely predictable nor readily described, even after completion.

4.4

The rise of the manufactured component

Despite the vociferous criticism towards post-war industrialised building, the failings of modernism's idealised union of architecture and manufacturing industry have not led to a demise in the industrialisation of construction generally. The industry continues to use a myriad range of manufactured products reflecting construction's adoption of a 'systems' approach to building production. This condition is not limited to specific building typologies or aesthetics, but is evident across a broad spectrum of construction activity. Notions of mechanisation, prefabrication and standardisation are all instrumental in defining the various characters of system building, as are models of organisation whose origins lay within factory-based concerns.

The systems approach relies on a widely held assumption that previous architectural practice has created a common resource of established technical knowledge: tried and tested solutions that form a reference for application to similar concerns and conditions. As well as a record of established technique this resource is also the depository for more subliminal observations, such as the prejudices of other participants towards particular methods, solutions and technological approaches. The value of this resource lies within its ability to enable designers to innovate not through blind invention but rather through the customisation

---

[7]Day, A. (1997). *Digital Building*. Oxford: Laxtons, p. 65.

of previous experiment; this highlights any continuity between different projects and forms a basis both for the industrialisation of the process and the professional status of those involved. As a practitioner, assuming a role within this 'genealogy of construction' distinguishes one from the similarly intentioned amateur.

Designers rarely obtain solutions based solely on an isolated analysis of a problem or required condition alone, but routinely and habitually refer to technical precedent within the design process.[8] In this manner previous tectonic characteristics of architecture are often categorised and judged in a manner relating to functional performance, reflecting the imperative to ensure a continuum of satisfactory building behaviour between successive projects. Recently this concern has been further prioritised using the notion of the 'performance concept', where the building is described primarily through its required performance in use as derived from client and user requirements alone; the technical characteristics of the building, its materials, components and junctions, are not prescribed, becoming a subordinate description in the hierarchy of project documentation.[9] The performance concept seeks to protect clients from the risks associated with commissioning unique entities, and inevitably leads to an increase in the use of standardised solutions, often without contributing further to the body of prior knowledge that those solutions themselves relied upon.

Existing systems approaches

The systems approach to building has previously been divided into two forms: process led and component based, each of which seeks to reconcile the conflict between differing industries for mutual benefit. Recent practices display less familiarity with any single approach, but their definitions remain useful in understanding the role of manufacturing within construction.[10]

The process approach is derived primarily from contractor's interests in the efficient organisation of site-based building activity, the nature of which is often craft-based and

[8]Pye, D. (1978). *The Nature and Aesthetics of Design*. Reprint 1988. London: Herbert Press, p. 59.
[9]Sebestyén, G. (1998). *Construction: Craft to Industry*. London: E. & F.N. Spon, p. 59.
[10]Groák, S. (1992). *The Idea of Building*. London: E. & F.N. Spon, p. 131.

traditional to a large extent. By standardising building forms, the construction process seeks to reduce variations within any building type, offering benefits in productivity that increase exponentially with quantity. Although production remains focused on the building site, process based systems also utilise prefabrication to some extent, and seek to reduce in-situ activity by making use of the standardisation exhibited by many building products, such as the room-based dimensions of timber and plaster panel products. Many of the practices of large volume housebuilders continue to be indicative of this approach.

Different motives are apparent in component-based systems, which ensure that building design will be directly related to the characteristics of a specific manufactured product or system. This approach is strongly influenced by factory-based manufacturers, but also has a direct effect on the nature of site-based construction processes. The use of component-based systems often reduces the amount of traditional building activity, such as cutting, joining and wet trades, and substitutes it with a simplified assembly process which is much less dependent on the skill of the workforce or the physical conditions of the site. Brought to its conclusion, this approach tends towards even the automation of the construction process itself.[11]

Separation

Although there has been an increase in the variety of industrialised techniques and materials exhibited within building production generally, this has not yet equated to a corresponding expansion in the palette of techniques with which architects and designers are familiar or utilise directly. In many instances design is inevitably reduced to the specification of components and systems originally conceived without reference to the uniqueness of the project, either as a technological or social construct. The level of uniqueness inherent to any individual building project necessitates a high level of innovation, which often appears incompatible with present client demands for reliability and economy. As a consequence, solutions are routinely sought within the many manufactured products and systems whose characteristics are already established and tested. The availability of 'ready-made' solutions has found building designers effectively sharing

[11]Sebestyén, G. (1998). *Construction: Craft to Industry*. London: E. & F.N. Spon, p. 180.

a

b

c

d

4.5a, b, c, d
Production sequence for a
washer. Reproduced from W.
Wilson Burden (1944). *Broaches
and Broaching*. New York:
Broaching Tool Institute.

authorship with the designers of manufactured systems. In some instances, this can prove a valid means to expand the scope of manufactured products within construction, but an uncritical reliance on this approach also serves to further distance architects from the process of manufacture itself.

This reduction in the architect's influence in the design of building components and systems is also a product of the differing nature of factory-based and site-based environments. The manufacture of products by mechanised means has been highly dependent on a unique production environment designed solely to produce one item alone. This created a significant obstacle to designers from disciplines outside of manufacturing who sought to utilise its techniques. Within the building industry designers were not familiar with the many fabrication techniques involved and, in addition, the high cost of mechanised production was prohibitive to the production of the relatively small batches of components that individual building projects often required. Predictability of demand was essential to factory-based firms whose significant capital investment cannot be recovered without many years of successful operation; yet this runs counter to the manner in which the building industry has continually attempted to re-configure its primary means of production by 'hiring and firing' the workforce. Understandably, manufacturing industry has been wary of incorporating the cyclical nature of the building industry into its long-term economic planning. Reflecting these conditions, building products are often manufactured only in response to constant demand and known needs of building performance.

A post-script to modernism

The icons of manufacturing that inspired Gropius and Le Corbusier were products of an organisational strategy that, almost by definition, separated the realms of art and technology that they so passionately sought to combine. This separation, characteristic of the 'loss of aura' associated with mechanised reproduction, stemmed from the inflexibility of mechanised techniques, which were not compatible with interdisciplinary endeavours that required unique solutions to one-off problems.

What is ironic about modernism's fascination with mass production is that its most prolific period of experimentation was at the very time when manufacturing industry sought to develop a new generation of techniques that would challenge the reliance on standardisation in which the separation of art and technology was rooted. While standardisation was the reality of mass production, it obscured a different and quite contrary goal of industry since its first beginnings, that of flexibility.

Mass production had offered new, complex and affordable products to new sections of the population that it itself had created. Unease amongst industrialists grew, however, as the world became increasingly littered with products for which demand proved unpredictable; it soon became apparent that mechanisation had created a strong, yet brittle, resource whose economic rewards could only be accessed through a strategy of increasing financial risk. As manufactured products proliferated, the success and longevity of mechanised enterprises became increasingly vulnerable to the whims of fashion and increasing world-wide competition.

Industry's ideal remedy for this condition would take the form of a universal manufacturing facility, which could be applied to the manufacture of a range of unique and customised products. In many mass-production enterprises flexibility had only been apparent in the actions of the workforce and, despite a range of differing approaches to mechanisation throughout the twentieth century, significant breakthroughs in flexibility were to be achieved only with the emergence of the digital computer as a reliable and affordable resource. The control of machines and assembly processes using computers suggested the manufacture of a range of differing goods in response to continual changes in demand, whilst also addressing the increasing problem of industrial unrest which had become endemic to mass production. The development and subsequent dissemination of computer control has proved to be one of the most significant technological events this century; although it influences virtually all industrial activity its possibilities are even now far from reaching their conclusion.

4.6
Despite vast sums spent on customer research and public relations, the Ford Edsel found no popularity with American car buyers. Only 35 000 Edsels were purchased in the first six months of production. Credit: Science and Society Picture Library.

5 The birth of automation: early developments in computer-aided manufacture

The development of numerically controlled machine tools

Since the eighteenth century, industry had consistently pursued a path of increasing division and specialisation. Modern mass production had evolved on the premise that the production of goods could be more efficiently achieved by fragmenting assembly into restricted and manageable domains. Workers needed simply to master one specific task rather than considering the totality of assembly.

Manufacturing, whose development was often led by the automobile industry, had focused its efforts on the mechanisation of repetitive operations. The basic machine tools developed in the eighteenth and nineteenth centuries, such as the lathe and milling machine, had to a large extent been modified into 'self-acting' machines that produced components without reliance on the 'hands-on' skill of the machinist. The operator's role was now one of supervision, the mechanics of the machine substituting for the manual actions of acquired skill. Even where manufacture could not be mechanised, as in many assembly operations, the actions of the workers were regulated by the motorised constancy of the production line.

However, not all sectors of manufacturing had been able to utilise mechanised techniques so readily. The aerospace industry maintained facilities that often still relied on a variety of conventional machine tools, each one being used to make a range of differing components. This approach reflects the much smaller quantities and greater complexity of products required, many of which were still in their design phase as production commenced. This placed a premium on flexibility, which was provided in the most part by the ability of skilled operators to use universal machines to make accurate parts in small batches. The resulting higher unit cost of the products was an inevitable result of this organisation, itself necessitated by the nature of the products it produced.

In a scenario reminiscent of the days of Eli Whitney, and the development of Interchangeable Manufacture in the nineteenth century, the 1950s was to see a period of concentrated development in the emerging field of computer controlled machine tools, predominantly within the United States. Despite an acknowledgement within industry for the need to develop more flexible manufacturing techniques, the impetus and financial backing in the development of computer control came once again from the US military. The extraordinary commitment of the government in the development of this technology circumvented both the conventional wisdom and limits of industrial economics, and was fuelled by a complex combination of technological, social and political factors. The story of its development is a testament not simply to the role of technology as a 'neutral' agent of change, but to the presence of other controlling aspects within industry and society that seek to create and exploit it. The subsequent dissemination of this technology, from the military sector into a broad range of manufacturing sectors, has caused profound changes in the organisation and capabilities of industrial production since that time.

Background

Developments in the computer-control of machine tools were to a large extent dependent on preceding research in both information processing and electronic control systems. Not surprisingly, many of the significant achievements in the development of computing reveal the clear influence and inspiration of Babbage's work to later researchers.[1]

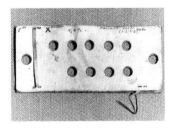

5.1
A punched card developed by Babbage for his Analytical Engines. Credit: Science and Society Picture Library.

In 1936 at Cambridge, Alan Turing undertook the first general analysis to understand how computing takes place. His attempt to formalise 'the class of all effective procedures' was demonstrated theoretically in the operation of his Turing Machine, which became fundamental to modern computational theory.

[1]Hyman, A. (1982). *Charles Babbage: The Pioneer of the Computer.* Oxford: Oxford University Press. p. 254.

Limitations in the practical capabilities of computers were only to be realistically addressed with the availability of electro-mechanical and semiconductor components. One of the first solid-state computers was the Automatic Sequence Controlled Calculator designed by Howard Aitken, and built by IBM during the war making extensive use of transistors. Further development steps were undertaken by Jon von Neumann in his Electronic Discrete Variable Computer (EDVAC) of 1945, which was the first electronic computer to execute a program stored within its own memory. Despite having no internal program storage capability it still found applications in mathematics, meteorology and hydrodynamics.

At a similar time, John Mauchley and J. Presper Eckert developed the Electronic Numerical Integrator and Computer (ENIAC) at the University of Pennsylvania. Developed with the support of the US Air Force it was used for a variety of military applications. Mauchley and Eckert then developed the Universal Automatic Computer (UNIVAC), which was the first to be produced commercially for a wide range of applications. Parallel developments in control technology at MIT, including the servo-motor, provided the possible means by which electronically stored data could be used to actuate a machine, whilst improvements in magnetic storage media allowed the storage of pre-prepared programs.

5.2
Experimental magnetic drum store, 1950. Credit: Science and Society Picture Library.

Replication

The earliest experiments in the computer-control of machine tools concentrated research on replicating the machinist's skills and actions. In the record/playback system the machinist would first make a part on a modified machine tool, allowing each of the subsequent movements and settings of the machine to be recorded on magnetic tape. This enabled the process to then be subsequently repeated in a consistent and repeatable manner, substituting the machinist for a lesser skilled operator.

The design of the record/playback system acknowledges both longstanding craft skills and the body of tacit knowledge that the machinist invests in the production of any particular part. The choice of machining speeds and the order of machining operations are both examples of the experience that is indirectly recorded on the tape. David Noble, charting the development of computer-controlled machine tools, has noted how this aspect of record/playback, rather than endearing it to the decision makers of industry, contributed to its failure to be promoted commercially. Despite its capability to automate, record/playback did not reduce the potential power that a skilled workforce had over its management, as key decisions affecting the success of production remained at the machinist's fingertips. This left management distanced from the production process and unable to enact Taylorist strategies upon the workforce. The system also had technical limitations, as it did not offer the capability to machine some complex three-dimensional surfaces, which still remained beyond the possibilities of manual control. As a result, record/playback, despite being developed into a working system by General Electric in 1947, was never utilised by manufacturing industry to any great extent.[2]

At a similar time, John Parsons, a US airforce subcontractor in Michigan, commenced a parallel investigation. The Parsons Corporation had successfully tendered for the production of rotor blades for Sikorski and Bell helicopters, and in order to check the geometry of the blades contour patterns were provided. In addition to these, punched cards were supplied to generate co-ordinates of the blade surface. Problems became apparent when processing of the punched cards revealed an unacceptable discrepancy between the geometry of the patterns and the required airfoil. To remedy this Parsons used the punched cards to generate more complete co-ordinate tables of the blade geometry, which he undertook originally using a tabulating machine and later a digital computer. This co-ordinate information was then used to produce patterns by reading out the numbers to the dual operators of a milling machine, each controlling the x or y-axis of the cutter path. Although a protracted process, the accuracy of the patterns was impressive, and Parsons was quick to see the possibilities of controlling a machine tool directly from numeric data.[3]

Having received support from the US Air Force, whose increasingly complex designs were close to the limits of available machining technology, Parsons subcontracted the research program to the servomechanism lab at MIT. In 1952 the first 'numerically controlled' machine tool was

---

[2]Noble, D.F. (1985). *Social Choice in Machine Design: The Case of Automatically Controlled Machine Tools from the Social Shaping of Technology* (D. MacKenzie and J. Wajcman, eds) Milton Keynes: Open University, p. 109.

[3]See Kochan, D. (1986). *CAM: Developments in Computer Integrated Manufacturing*. Berlin: Springer-Verlag, p. 5.

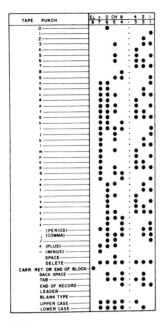

5.3
EIA Standard RS-244 perforated tape for numerical control systems. Credit: Science and Society Picture Library.

unveiled; based on a conventional milling machine it differed fundamentally from the record/playback machines demonstrated earlier. Whilst record/playback relied on data that only the actions of a trained machinist could supply, the numerical control system used a computer program to control the machine's movements. This made use of a new programming language, and moved control of the production process away from the shop floor and toward the managerial realm. That this transition was perceived as a benefit of the system is beyond doubt, as Noble's research shows, but the nature of numerical control was primarily a direct reflection of the requirement to machine complex three-dimensional parts. Noble notes how the mixed motives and multi-determinism in numerical control development caused problems during the dissemination of the technology, which stemmed in part from false assumptions by management and system developers as to the further relevance of machinists' skill.

Subsidy

The transfer of numerical control into the aerospace industry was of paramount importance to the US Air Force, but the high cost of the technology was prohibitive to all but the largest firms. In an act indicative of the charged political climate of the time, the US Air Force undertook to supply over a hundred numerical control machine tools to its key subcontractors. The overall cost to the US government was to be in excess of $60 million by the end of the decade, but despite this generous support the dissemination of the technology within industry was not without problems. In particular, the Air Force's insistence that a universal programming language should be used which, although having the benefit of standardisation, was to increase the costs of the system and the complexity of its use. The system's reliance on programming offered many opportunities in the machining of complex surface geometry, as it enabled its description by unambiguous mathematical means, but the reality of machining parts from real materials still required important contributions from machinists, whose experience could not be translated into the program so readily. In addition, unions had been excluded from discussions about the implications of the technology from the outset, and management's attempts to downgrade some machinists' posts inevitably provoked a negative reaction from the workforce.

The success of the US Air Force in developing and transferring numerical control technology to its suppliers was bought at great economic cost. No doubt, this was seen to be entirely justified by government given both the political climate and the nature of the products it would help create. However, without subsidies for other sectors of industry, the further dissemination of numerical control away from the military was to be at a much lesser pace, and its application and evolution within industry over the next decade was almost exclusively within large firms.

Distribution of control

The possibilities offered by continuing developments within computing, coupled with longstanding ambitions for a fully flexible manufacturing facility, initiated constant research into the broader applications of numerical control within large manufacturing firms. NC was now being used not only to automate single machine tools but to organise the entire manufacturing process, including the handling of raw materials and parts facilitated by parallel advances in industrial robotics.

Large firms began utilising a number of networked NC machine tools to automate the manufacture of a 'family' of differing components similar in their scale and application. Part programs were fed from a centralised computing source to specific machines in sequence; changes in material, tooling and part geometry were all accommodated without the necessity for manual intervention. These systems, first operational in the late 1960s, have become known as flexible manufacturing systems (FMS), and have increased in their sophistication and versatility since that time. The primary aim of system developers has been to increase still further the information processing capabilities of the manufacturing system to integrate all flows of information, material and energy throughout the process. These goals underpin the concept of computer-integrated manufacturing (CIM) currently exhibited by larger firms.

Exceptions to the rule

The rapid development that NC initiated for large firms has often obscured the diversity of the world's changing manufacturing resource, effectively substituting the clichéd image of the assembly line for one of the seemingly uninhabited machine halls of automation. Yet any industry operates at an extraordinary variety of scales and levels of technological evolution simultaneously. Its products can represent the output not just of the 'state of the art' condition but, rather more likely, the entire spectrum of available techniques, many of which will be typical of previous operational limits. Differing styles of management and organisation will exist, and specific practices often manage to endure long after alternative 'best practices' have been widely demonstrated. There is a continuing risk that when we describe the specialist techniques of an industry that those techniques alone are falsely taken to be the industry itself. This is even more apparent when we describe emerging developments, which do not necessarily have the precedent of previous use or classification.

In reality, industry functions at differing levels of efficiency extreme in their variation. Methods and management achieve longevity almost in spite of themselves, and successful ventures are as likely to be built upon a condition of chance as they are from consideration of market needs or technological innovation. There is, it would seem, a poetic and politic of industrialised practice where the firm is individualised by the conditions it operates in and its response to them. This is perhaps the most endearing aspect of these endeavours, and one that emphasises the importance of communication and technology transfer between different individualised concerns.

Up to now, the selected developments within manufacturing that have been presented in this volume mirror the approach of many industrial histories. They highlight the most distinctive and recognisable characteristics of the industry, those whose contributions are among the most iconic within our society. The steam engine and the automobile are undoubtedly two of the most significant inventions of our recent history, the first providing the power for industrialisation, the second a defining product of that industrialisation. A study of these products reveals some of the differing limits of the industry's capabilities at each particular time. In the case of Watts' improved steam engine it was the limit of precision that is so aptly demonstrated, whilst for the automobile it is the totality of the assembly line that forms such an enduring image. Yet what is much harder to assimilate is the diversity of available technology functioning at any one time and perhaps more importantly, which technology do we, as designers, genuinely have access to within any given period?

5.4
CNC lathe 1999. Credit:
Hardinge Machine Tools Ltd

Horizontal organisations

Sociologists, studying industry in Western Europe and America around the mid 1970s, observed that there were groups of small and medium-size enterprises, SMEs as they became known, that were successfully competing in a range of industries on a global scale. In addition to this, they operated within markets that previously only large corporations had exploited. These enterprises had not sought to adopt the organisational characteristics expounded by larger firms but, in passive opposition, had developed their own strategies based on a modern version of craft methods. These studies shed light on a scale of industry that is conspicuous by its absence from the majority of twentieth-century historical accounts, and runs counter to mass perception of the industry. The visibility of large volume production had eclipsed from many people's minds a significant proportion of an industry, which despite very different ways of working, was both innovative and successful. Of course, firms of this scale had always been in existence since the birth of industrialisation; the workshops of Watt and Clement in nineteenth-century England are obvious precedents. At that time, however, such workshops were typical and competition was between enterprises similar in scale and scope, so why was it now, with large volume manufactures ushering in the fully automated 'factory of the future' that such enterprises continued to appear and thrive?

Part of the answer was frequently found in the technology that industrial SMEs were exploiting. The first generation of NC machine tools had relied on a separate computing facility, but by now many machines possessed their own 'on-board' computer that allowed the machinist to program on the shop floor. The cost of these *computer* numerical control (CNC) machine tools had now reached a level that smaller ventures found affordable, however, unlike large firms SMEs were not using them as part of a large-scale fully automated facility but instead to supplement existing conventional machine tools and even manual methods of assembly.

In mechanised endeavours increases in efficiency were always contingent on the separation of conception and execution within the overall task. The management of such an organisation was strongly hierarchical, maintaining the distinct vertical divisions between the activities of management and machine operators.

In contrast to this, workers within these smaller hybridised industries included a key component of improvisation as part of their production activity. This might include the appraisal of a proposed design, a consideration of the methods that might realise it, and the concurrent revision of those decisions as production commenced and continued. In this context, automation had redefined the domains of skill within production, not necessarily through the value of technique but rather in the organisation that results when skilled operators seek solutions to unique fabrication challenges.

The success of these SMEs was in direct opposition to classical assertions that efficiency is directly related to divisions in labour. Their success aptly demonstrated that a 'horizontal' approach to the organisation of production was viable and efficient. Indeed, many characteristics of these firms seemed to offer distinct advantages over the Smithian alternative founded on division. This reintegration of conception and execution allowed individual firms to tackle a wider range of activities, and to find economic solutions to complex production challenges. The flexibility of the firm is derived not just from the fact that workers can use their own decisions as the basis to modify their own practice, but also in the utilisation of computer-aided manufacturing technology to ensure productivity within smaller volumes of production.[4]

CNC is now a generic strategy that may be applied to many machining, forming and manufacturing processes. Laser cutting machines, routers for woodworking applications, and sheet metal punching machines are some of the many applications that have been added to the first generation of milling and turning machines. The appearance of this technology in the workshops of craft-based enterprises marks a significant shift away from the restricted domains it had originated within. The very groups of skilled workers that numerical control had threatened to disenfranchise were now the same that used it to supplement their own apprenticeships and skills. Given the appropriate conditions, it seems computer-aided manufacturing did not initiate an inevitable decline in skills or in the value placed upon them, but created a new context for their use and development. This context emerges from the value of the tool as a means to achieve innovation and complexity, but was then furthered by the reappraisal of existing forms of organisation that the techniques themselves support. This contextual approach is the means by which what was originally shocking becomes palpable.

5.5
Manual and automated techniques combined at Ehlert GmbH.

[4]Sabel, C.F. (1995). *Turning the Page in Industrial Districts from Small and Medium-Sized Enterprises* (C.F. Sabel and A. Bagnasco, eds) London: Pinter, p. 137.

6 The image beguiled: the CAD/CAM era

Reliance

One of the most significant factors in the recent development and dissemination of computer-aided manufacturing has been the widespread adoption of computer-aided design techniques by engineering and architectural practice. The computer control of machine tools has been possible for almost four decades, yet, the long-awaited transfer of these techniques, from the specialist domains of their origin into the practice of design, was contingent on the re-classification of existing conventions of representation into digital standards.

Even after the adoption of CAM by smaller enterprises many designers, particularly those with a non-engineering background, were still distanced from use of the technology by the need to be conversant in specialist programming languages. However, in the last decade this requirement has been virtually eliminated by the emergence of software applications that compile the necessary program code from a virtual representation of the required object. This capability has effectively transformed programming into a visual activity, and has greatly increased the ease with which many automated fabrication processes may be accessed and interfaced, particularly for complex three-dimensional applications.

This graphical approach has also given rise to radically new methods of fabrication, which have moved the manufacturing process away from the industrial environment and into the studio or office. Solid Freeform Fabrication, or Rapid Prototyping as it is also widely known, has been developed specifically to replicate virtual prototypes in physical reality, and uses a 'solid model' CAD file as its instruction and data source. The methods of fabrication these machines display owe little to traditional machining precedents, which remove material in a subtractive manner, and instead accumulate form through the controlled deposition of material layer upon layer

Figures 6.1
Layer visualisations for an object built using fused deposition modelling (FDM®) – a solid freeform fabrication process – which is explored within the project by Sixteen*(makers) later in this volume.

in an additive manner. Intriguingly, this mode of manufacture allows the fabrication of three-dimensional forms so complex that they would be impossible to realise by conventional methods, and displays little reliance upon orthographic conventions. This partnership of representation and automation has allowed objects to be physically realised in days or hours from their conception, in some instances.

Solid Freeform Fabrication and other 'desktop' machining processes have provided a further generative and expressive function for computer modelling and visualisation within design practice. The role of CAD in these emerging translations from virtual to real is an increasingly active one, characterising the interface between design and manufacture through its own techniques and nature. Any ambitions we have to exploit the capabilities of CAM will, to a large extent, remain dependent on our ability to develop a critical approach to associated computer representation. As the range of 'tools and techniques' found in current CAD packages proliferates into endless encounters with lines, planes, surfaces and solids, it has become a resource that has become increasingly hard to assess. It is also apparent that the changing environment of virtual workspace is making comparisons with manual drafting techniques increasingly irrelevant. However, for all these innovations, much of their current application continues to follow patterns of use that remain normative and entirely conventional.

'Aura'

At present, many practitioners exhibit contrary attitudes to CAD and its associated techniques. Whilst design practice appears to be increasingly dependent on CAD its value within the design process is still only partially understood, and its universal adoption within commercial practice seems to have occurred in spite of a wealth of criticism regarding its techniques and their perceived implications on design. This has suggested to some that the considerable industry investment in this technology may be driven by additional subjective motives, independent of the value of the techniques themselves.

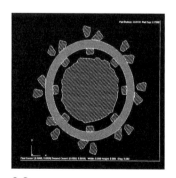

6.2

One such argument is that of Kathryn Henderson who, following Walter Benjamin, suggests that the 'aura' of high technology has been instrumental in the dissemination of CAD throughout professional engineering practice. She considers the notion that new technology, perhaps in a manner comparable to an original work of art, exhibits an aura that derives not simply from its capabilities, but rather through the mystification and rarity that is associated with it, particularly in its emergent phase. The status that this bestows transforms the tools of high technology into 'symbolic tools', which are recognised more for the status they grant to those associated with them than for their capabilities and functions alone. The acquisition of this status is considered essential to the continued existence of the engineering profession itself, as is the mystification and codification that is associated with it.[1]

At this advanced stage of dissemination, it is unlikely that many designers would feel this explanation alone encompasses their motives, conscious or otherwise, in adopting the computer into their professional routine. The majority would more reasonably attribute their subscription to the professional necessity of demonstrating efficiency comparable with related disciplines, rather than admit to their seduction by a technology that many consider (at their own risk) benign and familiar. Although the capabilities of the computer have always formed an allegedly irresistible invitation, many designers had critically declined it until the relevance to their own situation was more evident. Even so, the later inevitability of subscription was finally determined by the completeness of its distribution and the standardisation of its communication, far more than any individual's vision of its application.

Image

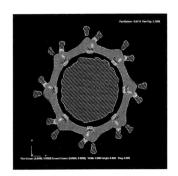

6.3

In his own discussion on the impact of the computer Daniel Willis has highlighted the futility and inappropriateness associated with attempts to make architectural practice more efficient. Willis considers that at the present time the 'logic of technique appears to exert a disproportionate control' over architectural practice, as it struggles to reconcile its own nature against increasing commercial and litigious concerns. The demands these concerns dictate are now, more than ever, based on a desire for certainty and predictability in building performance that traditional modes of architectural practice are becoming less able to give. Whilst not the cause of such trends, CAD use appears to be one of the few means by which they can be addressed, enabling practice to produce 'reliable', and often standardised, solutions which can be readily described before completion.[2]

[1]Henderson, K. (1999). On Line and On Paper. Cambridge, MIT Press, p. 189.
[2]Willis, D. (1999). *The Emerald City*. New York: Princeton, p. 279.

Similar pressures can also be seen to bear on the architectural rendering, which has expanded into a phenomenon strangely autonomous of the design proposal that it allegedly summarises. The need to produce a legible proposal, often within a time scale dictated by economic concerns rather than the eccentricities of the design process itself, has compelled practices to exploit the manipulative imagery of the computer in a manner comparable to that in which photography has been exploited to reconfigure the spatial and social reality of completed works. Here, the use of the computer has continued a tradition in which the myths of practice, and the proposed implications of its products on society, continue to be promoted within an idealised image. This enduring preoccupation excludes both maker and user from the reality of the proposed environment and the means of its realisation.[3]

Innovation and separation

Of course, the use of new technology does not necessarily reflect an innovative approach to design or production, unless it is explored within an application specific to its nature. The introduction of CAD in many large offices was preceded by a string of drawing systems whose goal had been to improve the efficiency of drawing production. Whether these systems have incorporated manual or electronic means, many have supported the increasing division of labour within the practice environment, becoming indicative of the dissociation of the design process into discrete phases of activity. These demarcations have created isolation both from the process of design and the techniques of construction for many, and comprise a considerable barrier to innovation within building design and research. It could be claimed that this argument is countered, to some extent, by the ability of the computer to empower the small and medium-sized office; despite this, it remains a fact that this empowerment rarely has a profound effect on the nature of production without an associated awareness of its physical implications upon building.

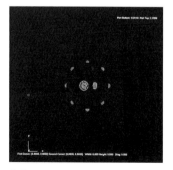

6.4

The practice of architecture, as any other design discipline, demands innovation in both the fields of design and technology. Gui Bonsiepe has defined the objective of design activity as the structuring of the interface between artefact and user and, as a consequence, innovation in design can be considered a form of socio-cultural innovation, becoming manifest in everyday social practice. This is in contrast to technological innovation

[3]Hill, J. (ed.) (1998). *Occupying Architecture: Between the Architect and the User*. London: Routledge, p. 135.

alone, which seeks only to develop new products, materials and processes.[4] The uniqueness and complexity of each building project necessitates simultaneous innovation in both fields, first making assessments about the viability of possible solutions, and then testing subsequent conclusions within any single project. In a situation where intuition and fact must continually revise each other, neither design nor technology can be separated; any attempt to extricate one from the other results either in technocracy, or a design based purely on aesthetic judgements. Increasingly, many practices appear to attempt to transfer design activity from one project to form the basis of technical innovation for another – separating related and contingent concerns in the process.

Even when these concerns are identified, the problem is misconceived if we seek to apportion blame upon the nature of CAD itself. Take, for instance, the concern of practitioners brought up with the techniques of manual drafting echoed by Daniel Willis, who claims that CAD has an inherent tendency to remove us from the real materials and methods of construction.[5] He considers that whilst both techniques share a common 'instrumental logic', a significant difference lies within the variation of speed and the degree of tactility that remain with manual techniques – these become the means by which we can partially mirror the actual techniques of construction. Willis argues that such contemplation is less likely with CAD, and that building shapes are too often 'disembodied abstractions'. It is clear that many architects have developed an empathy with the process of sketching and drawing which is certainly analogous to some aspects of making. The extraordinary sketches of Carlo Scarpa are a classic case in point, where his drawings seem to generate detail and form through an accumulation of lines that echo the addition or subtraction of material from a critical junction.[6] However, this analogy is not universal, and many sketches, despite displaying a complexity and enigma that alludes to some other technique, do not necessarily form an ontological link to the techniques that are employed in its realisation. Perhaps the point can be further explored by considering which techniques we seek to enlist through our act of drawing?

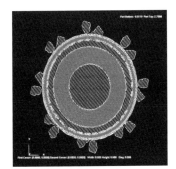

6.5

In a traditional example, say the construction of a timber joint, the manual process of drawing can also be seen as

[4]Bonsiepe, G. (1999). *Interface – An Approach to Design*. Translation. Maastricht: Jan van Eyck Akadamie, pp. 37–41.
[5]Willis, D. *The Emerald City*. New York: Princeton, p. 280.
[6]Groák, S. (1992). *The Idea of Building: Thought and Action in the Design and Production of Building*. London: E. & F.N. Spon, p. 151.

a simulation of the carpenter's first steps, as a tangible similarity exists between the marks of the drawing and the marking out of the material. These continue to have meaning throughout the joint's construction, as they govern the various operations of cutting and chiselling that they define. However, not all processes bear such direct comparison. Even ancient techniques of casting, forging or pressing bring about changes in the shape of materials less readily represented by conventional drawing, such as plastic deformation and changes of phase. Certainly, when we start to consider the techniques of CAM, machines operating under computer control are instructed by data that is presented visually primarily for our ease of reading and authorship. As is becoming obvious, the palette of techniques that inform modern construction cannot always be instructed by traditional drawing methods, or by any single convention alone, regardless of whether traditional or emerging techniques are used.

In an attempt to refine our approach to changing conventions of representation we seem compelled to acknowledge a meaningful disjunction between representation and building; an often missed opportunity that Robin Evans identifies as the 'unpopular option' in our attempt to understand the translation from drawing to building. Among his careful observations, Evans suggests that the modality of conventional architectural drawing is an expression of the perceived equivalence of wall and paper, with the drawing acting as both surface and veil for authored intentions in a manner readily transferable into building. Its limitations, however, stem from that very convenience, revealing inherent constraints in the nature of what may be represented and ultimately realised.[7]

By means of comparison, let us consider the solid model, a computer visualisation that attempts to describe the three-dimensional form and structure of an object without ambiguity. Though we may choose to capture and display this as an infinity of virtual images in various states of rendered conversion, its purpose and its nature are more meaningfully revealed through the allied medium of manufacture that enables its automated fabrication. Despite this, our reliance on the planar equivalence of projection is not completely removed, but remains simply as the means to navigate within these more dynamic representations.

---

[7]Evans, R. (1997). *Translations from Drawing to Building and Other Essays*. London: Architectural Association, p. 172.

Our sanction to use the virtual prototype to enlist computer-aided manufacturing lays intrinsically within its 'informating' qualities, transporting intentions autonomous of its surface qualities, and suggesting an explicit equivalence to physical actions as opposed to visible signs. This correspondence encourages us to specify our material proposals with a renewed definition, structuring a language that lessens our vulnerability to any falsehoods that linger within the drawing, or corruption by the status we have applied to it. Even the imminence of fabrication, and the increasing speed with which it might be achieved, suggests the latent power to render the drawing invisible and expended once the object is realised. It is that same speed of realisation that encourages us to repeat the translation again, increasing our understanding and fluency with every iteration.

Kathryn Henderson's many interviews with engineers using CAD has demonstrated that despite any changes in the nature of design practice suggested by the use of the computer, the sketch will continue to form a significant part of the visual culture of engineers.[8] She observes that engineering practice remains reliant on tacit knowledge, and identifies the importance of the sketch as an essential means to communicate that knowledge to other participants within the design process. This suggests that there will always be a place for 'low technology' in engineering design representation, as it is crucial in accessing the body of tacit knowledge on which the profession depends. This inevitably leads to the notion of mixed practices that utilise both high and low technology, not because they are in a transition from one to the other, but because both are necessary to each other.

What is becoming increasingly evident is the need to develop a new form of visual culture, one that derives as much from the computer model as from the manual sketch. Only when the aura of this technology has completely faded will we perhaps perceive more clearly the true capabilities of the tools at our disposal and their appropriate application. How this may come about, however, remains unclear; given the pace at which these technologies develop, can we ever envisage a time when we are so familiar with their capabilities and properties that they hold no mystery to us? With this in mind, we may have to resign ourselves that such a moment is unlikely to come, and that we must develop new practices that relish the prospect of continually exploring that which we do not fully master.

---

[8]Henderson, K. (1999). *On Line and On Paper*. MIT Press, p. 203.

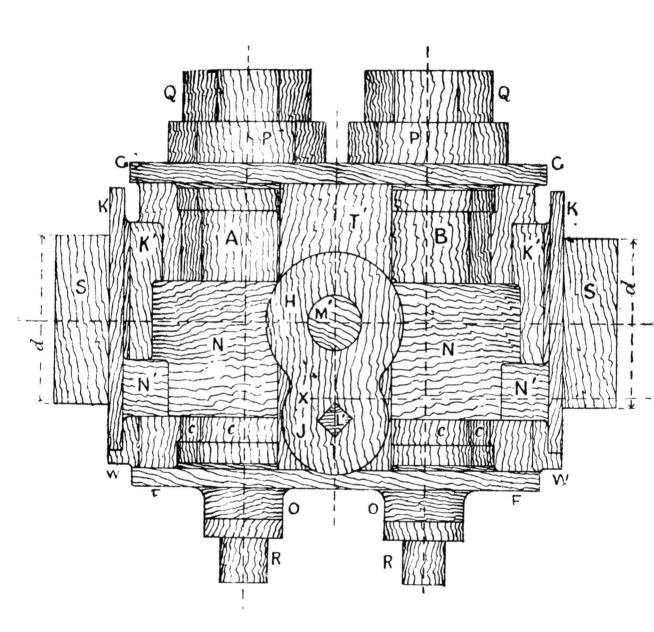

7 Reality and revelation

Apprenticeship

Designers and artists routinely call upon a body of craft knowledge; the ability to sketch with a pencil, make a model, or swiftly manipulate a geometric solid using a CAD program are all techniques requiring skill, dexterity and judgement. This is achieved not simply through the acquisition of technique in itself, but is also the embodiment of a tacit resource of experience that fettles our instincts in the search for what is appropriate, useful and enjoyable.

Traditionally, in many craft-based disciplines, the acquisition of this familiarity would commence with a prolonged period of training, formalised within many professions and trades as an apprenticeship. Using the differing environments of the academy and the firm, knowledge, application and experience are acquired both within the relative isolation of the studio and the reality of the workshop or site – the latter taking place in the real-time of production. To embark upon an open-ended exercise of this nature requires personal commitment, and an acceptance that the act of learning a technique creates a necessary separation from the practice that these techniques support.[1] This is not to say that learning and practice are distinct activities, but rather that the act of learning limits the domains of practice that we can contribute to at that time. Traditional apprenticeships manage the transition between these domains rigidly, often through the ritualised repetition of key skills within a hierarchy of supervision.[2]

Pattern for an engine cylinder. Reproduced from J. Horner and P. Gates (1950). *Pattern Making for Engineers*. London: The Technical Press.

[1] Dormer, P. (1994). *The Art of the Maker: Skill and its Meaning in Art, Craft and Design*. London: Thames & Hudson, pp. 40–57. See also Groák, S. (1992). *The Idea of Building*. London: E. & F.N. Spon, pp. 163–169.
[2] Dormer, Peter. (1994). *The Art of The Maker*. London: Thames and Hudson, p. 56.

In *The Art of the Maker*, Peter Dormer considers how, when learning a craft, content and process are interdependent. The acquisition of practical knowledge is an empirical process, and the constitutive rules which make up such knowledge cannot be apprehended intellectually until we are able to practise them. Within this realm, discrimination and judgement evolve in relation to our experience, and conceptualisation becomes potentially more meaningful. However, the validity and rationale of any apprenticeship relies heavily upon assumptions of stability and continuity within the subsequent domain of practice. Within many industries, including construction, these assumptions serve to create a repertoire of established approaches, techniques and solutions influencing future design, production and criticism.

At the present time, continuity in industrialised activity is rarely as evident as in the past, as emerging techniques and technologies rapidly evolve into new protocols and practices. If continuity exists at all, it is increasingly likely to be identified within other fields and disciplines, whose activities may previously have appeared incompatible or irrelevant with our own. The escalation of technology transfer between differing industries, and the popularity of self-directed learning equally suggest that the repertoire of many disciplines is now being expanded, and refined, through the activities of the individual in partnership with technology, rather than subscription to an institutionalised knowledge base.

The demise of the formal apprenticeship is often lamented as a false economy, reflecting the difficulty of quantifying its true value to either industry or individual concerned, but it is also a further reflection on the lack of continuity within current industrialised practice. However, the remaining alternative of 'improvising' our apprenticeship is often a lottery. Learning the constitutive rules of any practice within industry has become problematic, particularly so with the fields of emerging digital media and computer-aided manufacture, as the polarised applications of CAD within design practice dictate that any exploration of the digital medium's tacit domain must take place within the more prescriptive demands of the commercial project, and to date these latter demands have been more widely met than those suggested by the constitutive rules of the medium itself.

Consequently, it seems essential that our experimentation with computer-aided manufacturing be contingent on a reappraisal of the present roles, knowledge and conventions that constitute the path to production.

Patterns

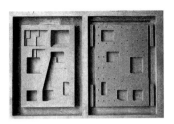

7.1
Pattern and mould production using CAM. Image and design: Nat Chard.

Design for manufacture typically involves many cycles of an iterative loop encompassing designer, engineer and maker in a collaborative partnership; one that is hopefully receptive to both conflict and reason. The creation of cast metal components is a classic example, and one whose path to production has been dramatically altered by CAM techniques, modifying practices and protocols that had previously existed unaltered for centuries.

Traditionally, the production of a casting has begun with the preparation of sketches or drawings of the component in question; often utilising established conventions such as plan and section. These representations, frequently incomplete or ambiguous as descriptions of form or surface, are then interpreted by a pattern maker at the foundry, a skilled joiner and carver, who creates a provisional pattern from wood or wax, say, for further appraisal. Specialist foundry and engineering input would be critical in discussions from this moment on, supplementing subjective debate with objective fact and informing further iterations of the design.

This process can be a time-consuming exercise, which is governed to a large extent by the physical actions of the pattern maker in shaping and joining the material; typically, several weeks may elapse before the pattern is put into production. However, the maker's role is not merely one of processing material, but is extended into a dynamic critique of the pattern throughout its making; a skilled craftsperson is able to enact judgement upon an object in the very instant of its making, as each action reveals a cumulative application of informed criticism. It is the same faculty that permits and forgives the ambiguity of representation in the first sketches, acknowledging that they are created as fragments in anticipation of a later tacit contribution from another. Characterising this traditional path is its reliance on the objectivity of the pattern itself, and the associated ability of its presence to initiate debate. This is of particular importance in the many instances of design where the desired outcome cannot be quantified or achieved through optimisation of performance alone.

Computer-aided manufacturing now comprises several alternatives to this process. Solid freeform fabrication techniques are ideally suited to the production of patterns for casting. Once the pattern is modelled as a virtual prototype its subsequent fabrication can take place as an automated procedure. Our previous reliance upon the pattern maker's skill appears entirely removed, and the object can be fabricated within the studio, originating in days, or even hours, from the moment the design was completed. How are we to comprehend this process, transformed out of all recognition from the traditional precedent, and which now seems to rely disproportionately on the availability of technological artifice in place of a previously tacit resource?

With automation, our understanding of representation is subject to an exacting test. Where previously drawings were interpreted and augmented with a degree of flexibility by a third critical party, decisions now made at the initial stage will be executed as unquestioned instructions. This suggests a need for the designer to be able to manipulate the solid model with all the dexterity and experience of the pattern maker, and to create a complete and 'knowing' statement where doubt and uncertainty may have dwelled before.

This automated translation also causes fundamental shifts in the locus of skills and criticism within the overall production process. The pattern maker's dexterity and experience appears a potential casualty of this transition, whilst the designer is presented with the intriguing, but seemingly mandatory, option of developing additional capacities through a more direct experience with form itself. This offers genuine possibilities for the dialogue between the virtual and the real domains, yet for this promise to be fulfilled, knowledge needs to be sought not from the digital domain alone, but also from those of the industries and crafts that it allegedly supplants. In relating these differing domains computer-aided manufacturing reconfigures a direct relationship between designer and object that remains structured by the act of 'making'.

In *The Nature and Art of Workmanship*, David Pye, mindful of the changing nature of making in the last century, offers us two distinct categories of workmanship.[3] The 'workmanship of risk' is that which we associate with traditional handcrafts, where the dexterity and experience of the maker remains the defining signature of the outcome. Even the highest levels of skill do not guarantee a successful

---

[3]Pye, D. (1968). *The Nature and Art of Workmanship*. Cambridge: Cambridge University Press.

outcome – carve away too much from an intricate joint and in an instant workmanship is compromised. The second category, the 'workmanship of certainty', is that which we associate primarily with mass-produced goods. High levels of repeatable precision characterise its products, fabricated with an array of mechanised machines and jigs that subjugate skill. Pye is clear to point out that both means of production are capable of producing meaningful results, as they are each similarly vulnerable to ineptitude. Is it now useful, or possible, to apply these categories to our pattern-making example, where automation and uniqueness exist in the same instant? As roles and categories go beyond previous precedent, has workmanship, or our understanding of it, also become stranded between the requirements of necessity and desire, certainty and risk?

In one sense, the making of the pattern appears to have made a very modern journey in circumscribing a path from risk to certainty, but this reductive conclusion encourages only a preoccupation with redundancy and suggests none of the expansive possibilities of computer-aided manufacturing. Although risk does not reside in the act of fabrication itself, it certainly remains evident in the activity of design that this technology is intended to serve, maintaining the opportunity to invest in the exploration of its use. The question remains, however, if we have only automated the craft of making how have we changed the craft of design? If a reductive analysis of the process does not reveal a sufficiently positive transition from existing modalities, then perhaps a consideration of its changing temporality may be more illustrative.

7.2
Pattern and mould production using CAM. Image and design: Nat Chard.

Revelation

Conventional craft workers receive constant feedback about the nature and outcome of their skilled actions through the full array of their senses; this intuitive connection is the means by which further improvisation is made. The instantaneous nature of this adaptive communication is fundamental to the skilled act, and its translation a necessary part of good workmanship. Originating from an automated enclosure, the material and surface of the 'rapid prototype' cannot engage with its authors in the same manner during its fabrication, yet once completed and liberated into their hands, the changing temporality of its production also creates an engagement that is both tactile and dynamic. Its creation remains, as is every act of making, a record of debate and an instigator of change.

When using these techniques both the act of drawing and the process of making are punctuated by a delay within the temporal landscape of design itself. Within this pause, expectant and energetic, our preconceptions are held in a state of suspended animation, before the reality of our actions and ideas is revealed. This creates a new rhythm to our work, yet pleasingly it is not one that is bounded by the constraints of a typical cause and effect relation. Though lacking the instantaneous improvisation of the pattern maker, this is not a world of mechanisation where skill is subjugated and where we are isolated from the means of production. With this understanding, the creation of objects using this technology seems to reside neither in a world of certainty nor of risk, but rather as a workmanship of revelation, where our skills, ideas and understanding are tested by their subsequent echo within a world of artifice.

This offers us a serious and hopeful means of speculation. Above all, we may find that through the materiality and tectonics of these curious physical constructs, we obtain a focus for our criticism that is not only shielded from the aura of emerging tools and techniques, but also acknowledges the social constructs within the technology of our built environment.

It seems we should now cease our comparisons of computer-aided manufacture with artisan activity, and consider instead our present relationship with the act of making, which is often as dysfunctional as it is complacent. In reality, very few designers could speak directly of the collaborative experience with the pattern maker. The designer's distancing from the maker commenced even before mechanisation; indeed the very notion of the designer is rooted in the earliest separation of conception and execution. Computer-aided manufacturing provides an opportunity to re-cast a relationship damaged long ago in creating a means of making concurrent with our design activity, allowing us to experience the actions and phenomena of manufacture more directly than mass production ever permitted, and yet avoiding the problematic notion of craft as a privilege. By direct association it reveals a resource of both traditional and emerging manufacturing methods, which combined allow us to speculate again as to our understanding of 'making'. Given time, its provision is unlikely to be either redundancy or isolation, but rather a means lacking since mechanisation for designers to learn dynamically about the presence of the objects whose authorship they claim.

8 The hybridisation of production

It has often been said by engineers that, as information levels rise, almost any sort of material can be adapted to any sort of use.[1]

Feasibility

The technology of CAM has realised the long-term goal of manufacturers and designers alike in reconciling uniqueness with economy. As mechanisation declines, the diversity of the products within the artisan workshop and the productivity of mass production seem at last to be mutually compatible. Now, with mass customisation comprising a practical approach instead of an idealised goal, the conception of an environment populated from unique entities seems a tangible proposition. As we consider the nature and production of such an environment many issues are raised, perhaps the most intriguing of which is the notion of its complexity.

Various definitions of complexity have been influential to the practice of architecture in its recent past. The enthusiasm with which these collective ideas have been embraced reflects complexity's dual ability to navigate away from the rationalism of previous eras and to develop further insight into the nature and behaviour of the built environment, and its population, in broad terms. CAM has already been instrumental in the realisation of architectural projects whose conceptual roots exhibit both a valuing of complexity as well as an imperative to bring such a condition into being. Frank Gehry's Guggenheim Museum in Bilbao is by no means the only architecture that has been generated and realised by utilising the capabilities of

[1]McLuhan, M. (1964). *Understanding Media: The Extensions of Man.* Reprint 1997. Cambridge: MIT Press, p. 351.

customisation which the computer has made feasible; Norman Foster Associate's glazed enclosures at Canary Wharf Underground station, Alvaro Siza's Portuguese pavilion at the Hannover Expo are other recent applications.

The growth of a new science of complexity has to a large extent overshadowed existing architectural thought that focused on the contradictory dialogue which exists between many disparate forms, creating an almost combative relationship between elements that retain their autonomy in spite of each other. The attainment of any equilibrium is neither through symbiosis or union, but rather through an almost personal reading of a landscape littered with symbols. Despite the ability of this approach to highlight the multiplicity around us, this definition describes a complexity that is both passive and static; alluding to the temporal in the sense that it identifies a moment suspended which is already past.[2]

In contrast to this is the realisation of complexity in nature, in which multiple simple systems interact to create complex and emergent behaviour. The continuing synthesis within these interactions identifies the creation of complexity upon a temporal landscape, across the topography of which emergent behaviour may be seen to develop and change. This characterises a dynamic domain, but one in which an overall condition may be more clearly discerned, and has sparked a cultural realisation that this underlying process can be replicated in the things we make and use.

The willingness to exploit both these and related ideas stems from a variety of motives. For some, complexity offers a vehicle that is capable of deriving and appropriating formal systems in a manner illustrative of the dynamics between environments and their occupants. This position sets a formidable challenge to normative modes of description, and requires a consideration of methods of spatial representation that has lagged behind an understanding of complexity itself. Cartesian conventions of representation have remained the dominant mode of description, whose reductive analysis struggles to describe the phenomenological possibilities between an environment and its users. Some tentative improvements to this approach are already visible within the modelling approach adopted by computational fluid dynamics. Derivations of these simulations may become a

[2]Lynn, G. (1995) Blobs in *Complexity: architecture, art, philosophy, The Journal of Philosophy and the Visual Arts*, special edition vol. 6, ed. Andrew Benjamin.

means by which issues of the site and user may be more usefully considered within the design process itself. Designs may be propagated from a variety of contextual influences, whose subsequent interaction is not solely through authored editing and compromise, but is instead one that transforms the process of design into what Steven Groák has termed a 'non-invasive procedure'.[3] This requires models and simulations whose generative component is not based solely on the continuity of physical laws, but which is influenced by the vacillations of environments, authors and users alike.

But what of form itself? These approaches do not necessarily imply an associated non-Cartesian approach to form in themselves, yet it is also clear that notions of complexity appear to have influenced the development of form away from the orthogonal and into non-linear and 'organic' modes of composition. This has created an inevitable association of these modes with computer-aided manufacture, as it remains one of the few technological means by which it might be realised. Whilst not intending to censure these experiments in any way, this single association alone threatens to marginalise CAM long before its broader capabilities are explored. This requires an understanding of CAM not as a technological determinant alone but as a social construct whose domain is contextual with design.

Many ideologies and representational modes implicate within their concerns methods and techniques that may be enlisted for translation into form. Ideology and technology have often formed coalitions whose associations have outlived their products. The passionate rhetoric of Morris or Gropius towards their specific technological position has only served to highlight how architectural design differs from other engineering design disciplines, where it is possible and necessary to identify an optimum production method for each product. Not only is it impossible for any single method to bring about realisation within architecture, even the range and combination of these methods will remain unique to specific projects and authors irrespective of universal performance requirements – the essence of architecture relies upon this synthesis.

Constraints

Technological endeavour is characterised and structured by the attainment of identifiable goals, and the first generation

---

[3]Groák, S. Project-related Research and Development in *Arups on Engineering* (D. Dunster, ed.) Ernst and Sohn.

of architecture to exploit CAM enlists technology to express very specific spatial conditions. In the work of Frank Gehry at Bilbao we witness the combative, though essentially static, superimposition of non-linear volumes, while in Foster's glazed curving enclosures it is the introduction of a measured variety within an otherwise still standardised territory that is exhibited. Although in the latter example, geometry may be more readily described in Platonic terms, both examples appear to reside within an essentially reductive approach, that differs from those previous only in the removal of traditional economic production constraints which CAM has facilitated. Despite this, the very existence of this work demonstrates that the established relationship between formal complexity and the feasibility of its realisation is now dramatically and irreversibly transformed by CAM, creating a revolutionary means of expression; perhaps unparalleled since the emergence of iron and steel as building material?

Constraints upon design activity may take radically different forms in this context. With the reduction of formal complexity as a constraint, and with the economics of variety and novelty re-written, our relationship to production becomes liberating once more. This may suggest to some that this technology may be applied 'at will', being able to support architecture as a generic strategy of production independent of ideology or aesthetic, but a conclusion of this nature must be avoided if this technology is to be fully understood. Any drift towards a generic can also imply neutrality, a term which can only be applied to any technology erroneously. The reduction of these constraints does not only provide opportunities in the development of form, but more significantly comprises the means by which the authored intentions of design find compatibility with the technological choices of production, realised within a mode of expression both formal and phenomenological.

Continuity

As the notion of complexity in architecture is reconsidered it appears its existing definitions do not encompass the range of activities or relations that comprise the production of the built environment. While this spectrum is itself considered complex, an awareness of its nature has not directly informed strategies of design which are influenced by complexity. Central to a re-appraisal of this condition is a further consideration of the temporality of the path to production. Manufacture should no longer be considered as a discrete phase of pre-determined activity, but should be conceived as a considered mediation between instructions, actions, objects and environment. If production is not defined in this manner the full scope of CAM's implications for the built environment risks being re-cast as the object of a fetishistic desire. The application of this technology at the scale of the built environment requires not only acquiring a familiarity with its capabilities as exhibited in its domains of origin, but also undertaking the task to consider its rules of engagement with current building production techniques and craft-based disciplines. This hybridisation can rarely be justified by efficiency alone, but instead should be considered a further element of design intent, where the multiplicity within any solution characterises a unique and individual approach to the realisation of the whole. Although using widely different means, both CAM and the building industry each continue to exhibit capabilities of flexibility and diversification that suggest an assured compatibility.

As we go beyond our preconceptions and learn more about the capabilities and realities of CAM, the many techniques we find at our behest seem rather more familiar than broad terms suggest: cutting, drilling, folding and layering – these are actions with which *homo faber* may relate through direct experience. The results of these actions may be discrete, but their combination remains an act of making reliant upon assembly, as additive and accumulative on-site as it is within the enclosure of the machine. As a result, the transfer and integration of CAM into construction requires an evaluation of existing practices and a need to reconcile these traditional methods with emerging means of production. Indeed, CAD/CAM technology may prove the vehicle through which existing methods are more fully utilised, as the critical selection associated with hybridisation becomes the mechanism of learning in our re-definition of the construction process itself.

In 1964 Marshall McLuhan concluded his book *Understanding Media* with the essay 'Automation: Learning a Living'. Extreme in his optimism, McLuhan saw automation as an irreversible break from mechanisation and its separation of operations, in which learning is liberated to become the principle form of production and consumption.[4]

[4]McLuhan, M. (1964). *Understanding Media: The Extensions of Man.* Reprint 1997. Cambridge: MIT Press, p. 351.

The very toil of man now becomes a kind of enlightenment. As unfallen Adam in the Garden of Eden was appointed the task of contemplation and naming of creatures, so with automation. We have now only to name and program a process or a product in order for it to be accomplished.[5]

At the heart of McLuhan's optimism is electrical energy, and the 'instant speed' it facilitates. Between the separation of electricity's generation and the subsequent application of its power McLuhan considers a fusion of energy, production, information and learning is inevitable.

Our new electric technology now extends the instant processing of knowledge by inter-relation that has long occurred within our central nervous system. It is that same speed that constitutes 'organic unity' and ends the mechanical age that had gone into high gear with Gutenberg. Automation brings in real 'mass production', not in terms of size, but of an instant inclusive embrace. Such is also the character of 'mass media'. They are an indication, not of the size of their audiences, but of the fact that everybody becomes involved in them at the same time. Thus commodity industries under automation share the same structural character of the entertainment industries in the degree that both approximate the condition of instant information. Automation affects not just production, but every phase of consumption and marketing; for the consumer becomes producer in the automation circuit, quite as much as the reader of the mosaic telegraph press makes his own news, or just *is* his own news.[6]

Branded both populist and naive at the time, the fallacy of McLuhan's wilder predictions obscure the considerable accuracy of many of his insights, whose value has taken time and events to appreciate.

One of McLuhan's fundamental distinctions between mechanisation and automation is the adaptive possibilities of the latter, facilitated by the electrical information network it resides within, yet the fusion of production and learning he foresaw was for some time hard to discern amongst modern activity. In the interim, automation was appropriated by large-scale industry as simply another form of mechanisation, indistinguishable from its predecessors by the majority of the workforces that maintained it. Only later, when the technology of automation reached the wider

---

[5] *Ibid*, p. 352.
[6] *Ibid*, p. 349.

audiences of the small and medium-size enterprise (SME) and the studio, has the fusion between information, production and learning become a familiar and tenable proposition.

Facilitated by the information highway of cyberspace and exploiting technologies old and new, McLuhan's automation is exactly that of the desktop rapid prototyping machine, able to transform even the most domestic of space into a site for the most sublime of fabrications. Perhaps these new forms of automation will come to alter our designations of space and use as surely as the TV attempted to turn the parlour into a cinema, re-inventing our expectations of the factory, the site and the studio?

Part II
Enlisting the manufacturing process

Reproduced from W. Wilson Burden (1944).
Broaches and Broaching. New York: Broaching Tool
Institute.

9 Shorting the automation circuit

Design: Sixteen*(makers)

Process: Fused deposition modelling (FDM®)

The effects of computer-aided manufacturing have often been defined simply as a hastening of the path to production, but in this site-based installation an attempt was made to re-configure that path itself, placing authors, objects and environment within the 'automation circuit'.

The work exists both as object, system and event, the overlap of which creates an opportunity to develop extensions to existing design practices through a direct relationship with the manufacturing medium. It also seeks the revision of our existing understanding of production by constructing an interactive relation between site, audience and representation that is activated by the manufacturing process itself.

Process and technology

The project is sited within a redundant observatory at University College London, whose existing mechanical roof provides an articulated threshold to the surrounding public quadrangle. It is here that our object resides, reviving a tradition of observation and collection long past.

The object itself embodies a series of requirements and influences, which are functional, sculptural and contemplative. Housed within its hollow interior is an array of light dependent sensors, each of which are selectively activated by the light refracting through an acrylic conduit whose angular position is governed by prevailing air currents. The internal cavity of the object is structured by the sensor array geometry, whilst its outer surface topography reveals a further desire to develop additional sculptural properties of that geometry by exploiting the capabilities of the solid

*Phil Ayres, Nick Callicott, Bob Sheil

freeform fabrication process. The final object exhibits many intricate internal features, and would be impossible to fabricate as a single component using conventional machining or handcraft. Its virtual description was constructed as a solid model using a CAD package, and converted into an .stl file prior to manufacture.

In use, the object's sensors record an informal 'thumbprint' of the prevailing physical conditions, one that is also inclusive to the more dynamic presence of passing spectators. The sensor signals are digitised using a STAMPII control processor interfaced to a PC, and a terminal program then collates and tabulates the data into a text script suitable for import back into a three-dimensional CAD application.

The transference of this environmental data into the modelling program suggests a number of subsequent avenues of research. If required it could be directly converted into form, providing a legible means of representation for some environmental analysis applications. In this instance, however, the intention is more generative and seeks to integrate the information with subsequent design proposals for the site. By combining the data with existing authored representations we can experiment with the 'weathering' properties of the data upon previously modelled form, adding or subtracting material in the search for specificity.

The approach does not constrain or prescribe our creativity, whose outcomes remain evident both in physical form and within the algorithms and architecture of the system itself, yet the underlying desire is to reconcile this authorship with the social, environmental and technological factors that comprise the reality of the site, and its means of production.

Fabrication of the fused deposition model was carried out at the Department of Medical-Physics, University College London with the assistance of Dr. Robin Richards.

FDM is a registered trademark of Stratasys products.

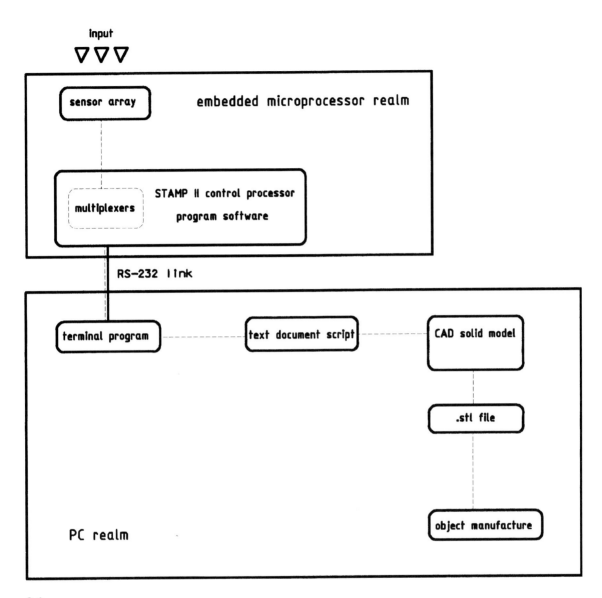

9.1
System diagram.

Figure 9.2a, b, c, d, e, f, g, h
The object in various stages of development. It was fabricated using the fused deposition modelling process. Note: The darker parts of the object in some images are support structures, generated automatically by the FDM machine's software to support overhanging structures during fabrication.

9.2a

9.2b

9.2c

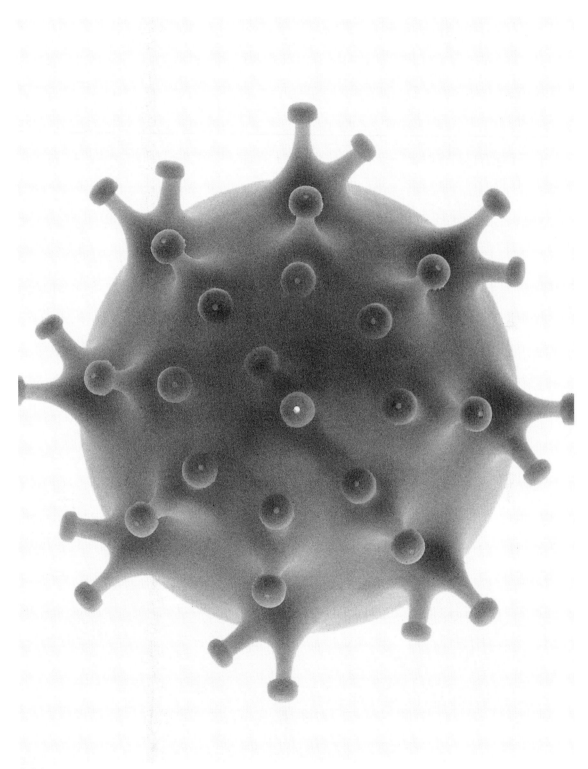

9.2d

9.2e

9.2f

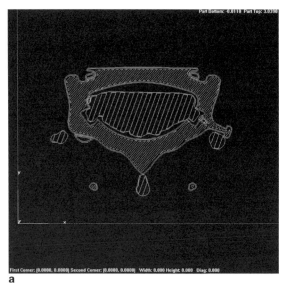

a

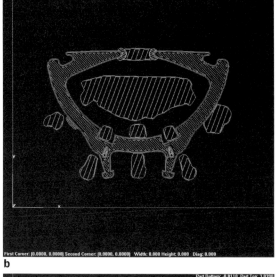

b

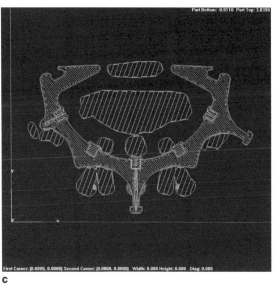

c

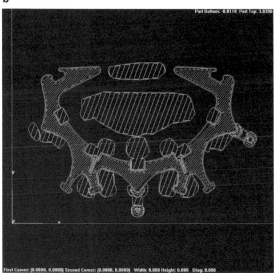

d

10.0a, b, c, d, e
Simulations of individual layers
prior to fabrication

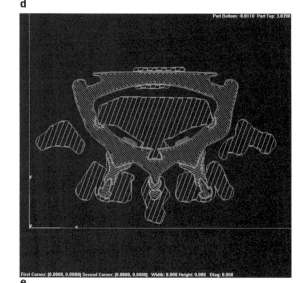

e

10 The myth of standardisation

Design: Sixteen*(makers) in collaboration with Kristina and Jürgen Ehlert
Process: CNC plasma cutting and CNC metal folding

The benefits of existing standard steel sections are many and varied. Available in an extensive range of sectional forms, they offer an economic means to realise many building structures and enclosures.

The language that these elements create has proved fascinating for many designers, and their lexicon of expression has developed a considerable fluency since their emergence in the nineteenth century. Their continued use is one of the few moments of continuity in current building practice, particularly in solutions for medium and large-scale structures. Designers have come to rely on their widespread availability, their known performance and a standard repertoire of techniques and details for their connection and assembly into a complete structure.

The prefabrication of a steel frame structure is effectively a series of operations, under factory conditions, that transform elements from a standardised to a unique condition before assembly on site. Although conventional, the fabrication required for this transformation represents a significant cost and time component when compared to the acquisition of the stock material alone. In many instances, the paradox of using a standardised form that then requires considerable customisation is self-evident.

Despite the benefits of standardisation, designers continue to seek structural elements unique to single projects, whose form is a direct result of their required performance, context and authorship. In this experimental collaboration, structural steel elements are explored as semi-automated one-offs in the search for a fully customised, yet economically feasible, building structure. The design is grounded in a desire to exploit the transition from planar to spatial, both at the level of the individual component and still

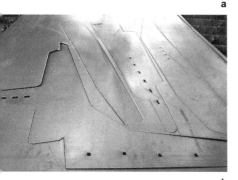

a

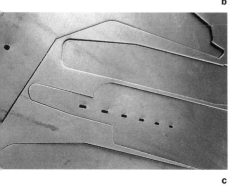

b

c

10.1a, b, c

further in their subsequent combination. Knowledge of the physical capabilities of the processes used was a prerequisite for both conception and execution.

Process

Individual components, and their combined assembly, were initially modelled on a standard CAD package, and their two-dimensional 'netshapes' exported as .dxf files. All holes and slots for future assembly are included, as are location marks for subsequent folding operations.

Using a dedicated software package specific to the CNC plasma-cutting process the .dxf files were converted into a numerical control part program. Elements were also manipulated visually to create a cutting layout that reduces waste. Engineering data, including the weight of elements, was also provided. Once compiled the part program is transferred by disk to the cutting machine, where the operator can make further modifications as required.

Subsequent folding operations were undertaken on a CNC press, which in this instance was programmed on the shop floor using its on-board controller. For more complex fabrications complete bending sequences can be programmed, and automated 'stops' assist the operator in locating the correct fold line.

The finished element is approximately 3 m in length, although the equipment used is capable of producing similar elements with a length of up to 6 m and with a thickness of 10 mm.

Fabrication was undertaken by Ehlert GmbH, Güsten, Germany. The firm has specialised in the production of one-off steel structures, vessels and machinery for the chemical and petroleum industries since its founding in 1936.

10.2a, b

a   b

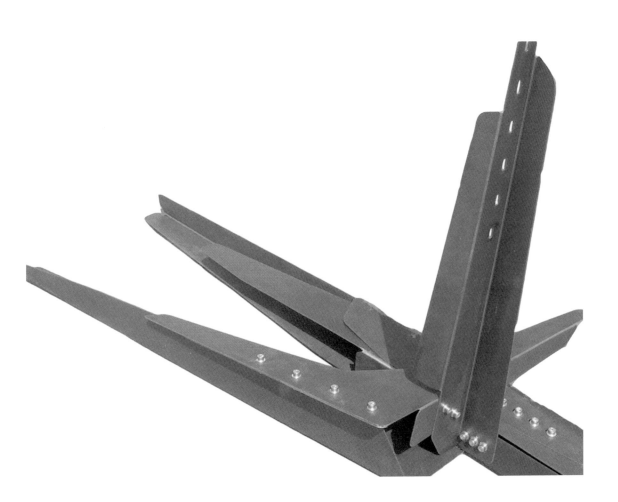

10.3a

10.3b

10.3c

11 Making Buildings

Design: Sixteen*(makers) in collaboration with Kristina and Jürgen Ehlert

A number of issues from the investigations in the preceding two chapters were developed and combined as part of Sixteen*(makers) contribution to the Crafts Council touring exhibition 'Making Buildings'. The installation is an exploration of boundaries real and virtual, responding to occupancy and use.

The exhibit seeks to create a network of objects in conversation with one another and with their local environment. It comprises a family of structural steel and plastic enclosure elements, which are configured in a number of permutations at differing locations. Housed within its folds and surfaces are a number of sensors that record the movement and proximity of those around it.

By networking to a micro-processor control system the information initiates a number of responses from the structure, each of which seek to create and sustain a changing environment both local to the exhibit, and virtually using the Internet. Local changes are accomplished using an integral lighting system and a series of physical actuators. As a result, those confronting the work experience the selective lighting of the structure and its surroundings - a blush, and the subtle movement of the secondary structure - a touch. Together they seek to create an environment unique to the moment that extends notions of building into an interactive condition.

The structural steel components were fabricated at Ehlert GmbH, Güsten, Germany.

---

* Phil Ayres, Nick Callicott, Chris Leung, Bob Sheil

The installation at the New
Art Gallery Walsall, 2001

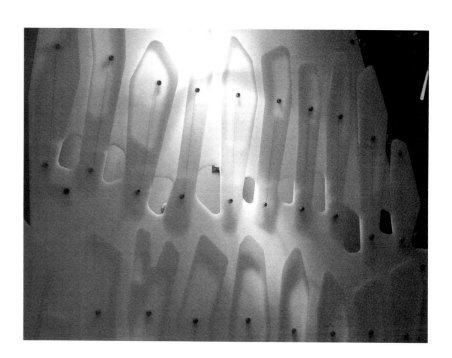

Part III
Technology

Reproduced from W. Wilson Burden (1944). *Broaches and Broaching*. New York: Broaching Tool Institute.

12 Computer numerical control

Introduction

'Numerical control' was the first electronic computer-aided manufacturing technique to be developed and emerged from US military research in the 1950s. The NC system was originally conceived to allow the automation of conventional machine tools within the aerospace industry, controlling milling machines and lathes. Initially a system for specialist applications, numerical control is now considered a generic strategy that may be applied to almost any manufacturing operation. Computer numerical control systems (as they are now more commonly known) can be found as part of a growing range of machine tools and fabrication equipment supporting cutting, machining and joining applications. NC systems allow numerical data stored within a computer to control each aspect of a machining operation. In many instances this involves guiding a cutting tool through a complex three-dimensional

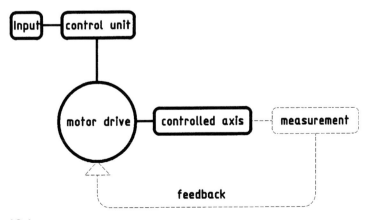

12.1
System diagram for a numerically controlled machine tool.

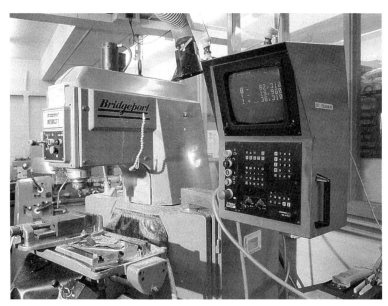

12.2
A three-axis CNC milling machine.

path, something that is often impractical when machines are operated manually.

Computer controlled machine tools have undergone a series of changes since their introduction. Despite significant improvements in the physical capabilities of machine tools, the pace of developments has been governed primarily by advances in the power and affordability of computers and associated storage media.

Stages in development

Numerical control
The first generation of numerically controlled machine tools received their instructions via punched or magnetic tape. This was prepared using early digital computers at a separate location, whilst the tape reader hardware was connected directly to the machine on the shop floor. NC systems executed the part program automatically, but did not provide operators with the means to edit or override machine operation. Any amendments would require the preparation of new tape.

Computer numerical control
Following advances in the miniaturisation and power of microprocessors in the 1970s, computer numerical control

machines began to incorporate on-board computers as an integral part of the machine. NC machines had simply executed instructions read from a pre-prepared tape, but *computer* numerical control machines were able to store the program electronically within the control unit. This enabled the writing and editing of programs by operators at the machine, making it far easier for machinists and designers to incorporate practical experience into the manufacturing process.

These features greatly simplified the programming process. Rather than create a program from scratch, operators were now able to use previously stored or standard routines as the basis for a new program. Part parameters could also be input simply in response to prompts from the machine. This method, which greatly reduces programming time, is often known as *conversational* programming.

The increasing power/cost ratio of programmable controllers is one of the most significant factors in bringing the early benefits of numerical control to small and medium-sized firms, enabling the use of CNC machines as 'stand alone' tools without the need for a costly centralised computing facility.

Direct and distributed numerical control

In large-scale applications, where a number of NC or CNC machine tools are operating, it has become common to network them to a single mini or mainframe computer. Part program data is stored centrally, and is sent to individual machines as dictated by the manufacturing process. DNC systems vary tremendously in their scale and complexity, and often form the basis of a fully automated manufacturing environment. The use of CNC machines in this system is often described as *distributed* numerical control, as the on-board computing capabilities of the tools share process control with the centralised computing facility.

The PC era

The control of machine tools requires fast real-time computing performance. The machine controller must co-ordinate each of the physical processes of fabrication, collecting feedback information and executing instructions both within a reasonable time and to the required accuracy. Even with the advent of the Personal Computer in the 1970s this real-time performance could still only be met with dedicated controllers designed and built specifically for the task.

At a similar time the use of CAD systems to describe components and compile part programs became more widespread, but as programs became larger their storage began to go beyond the means available within most on-board controllers. A PC appeared to offer a convenient and standardised means to manage this expanding information, particularly as many existing controllers had offered limited file management and editing facilities. However, despite the advantages of interfacing CNC machine tools to PCs, the specialist and unique nature of many machine controllers made the task problematic.

PCs were first successfully interfaced to CNC systems using serial interfaces, in much the same way as other peripherals were connected, but this resulted in serious limitations in performance for more complex applications. Current practice focuses on a hybrid approach, in which a dedicated machine controller is connected to a PC through an internal bus to achieve far greater speeds of data exchange. Such systems have the advantage of PC compatibility with the necessary high-speed capabilities of dedicated controllers.

Flexible manufacturing systems and computer-integrated manufacturing

From the outset manufacturers have sought to use CNC machine tools as the building blocks of complete manufacturing systems that are automated, flexible and intelligent.

Since the late 1960s flexible manufacturing systems have been used to automate the manufacture of a range of products with the minimum of human intervention. Each FMS exists as a fully networked structure comprising machining, material handling and system control elements. Flows of parts, material, tools and data are all synchronised, making extensive use of automated vehicles, conveyors and parts handling equipment. As a result individual CNC machining modules are able to produce a range of differing parts, as related raw material and program code are fed to machines in a pre-determined sequence. Although displaying far greater flexibility than mechanised production lines FMSs are often used to manufacture a family of parts similar in their material, scale and application, for example, the machining of differing engine blocks and other similar sized components.

FMSs comprise a number of manufacturing 'cells', each of which contains a computer, a CNC machine tool(s), parts handling equipment and waste removal. Cells may function as stand alone units, or be part of a larger integrated network.

Even greater levels of automation and flexibility can be achieved within the manufacturing environment by increasing still further the information processing capability of the system. At the extreme is the concept of *computer-integrated manufacturing*, a term which broadly describes the integration of each aspect of the process including design, planning, manufacture, distribution and management functions.

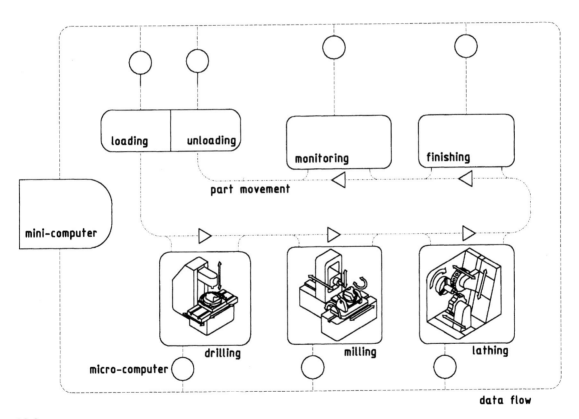

12.3
A typical FMS cell. Varying numbers of CNC machines may be employed.

Procedure	Dialogue initiation	Program block display

Traverse to tool-change position	TOOL CALL
	L
	
	L
		1 TOOL CALL 0 Z S 0,000
		2 L Z+50,000
		                           R0 F9999 M
		3 L X−40,000 Y−40,000
		                           R0 F9999 M05

Tool definition 1, programmed Stop and call-up of tool 1
Coarse-fine mill (4 flutes) Ø 20 mm
	TOOL DEF
	
	STOP
	
	TOOL CALL
		4 TOOL DEF 1 L ....
		                  R+10,000
		5 STOP
		                             M
		6 TOOL CALL 1 Z S 250,000

Positioning blocks for starting position, tangential approach and milling of contour, tangential departure from contour
 	
	L
	
	L
	
	L
	
	RND
	L
	
	CC
	C
	
	CC
	C
	
	L
	
	RND
	L
	
	CC
	C
	
	CC
	C
	
	L
	
	L
	
	CC
	C
	
	CC
	C
	
	L
	
	RND
	L
		7 L Z−20,000
		                           R0 F9999 M03
		8 L X−12,000 Y+60,000
		                           R0 F9999 M
		9 L X+20,000 Y+60,000
		                           RR F40 M
		10 RND R+5,000
		11 L X+50,000 Y+20,000
		                           RR F40 M
		12 CC X−10,000 Y+80,000
		13 C X+70,000 Y+51,715
		                    DR+ RR F40 M
		14 CC X+150,000 Y+80,000
		15 C X+90,000 Y+20,000
		                    DR+ RR F40 M
		16 L X+120,000 Y+20,000
		                           RR F40 M
		17 RND R+20,000
		18 L X+120,000 Y+60,000
		                           RR F40 M
		19 CC X+120,000 Y+90,000
		20 C X+118,266 Y+119,950
		                    DR− RR F40 M
		21 CC X+90,000 Y+130,000
		22 C X+90,000 Y+160,000
		                    DR+ RR F40 M
		23 L X+70,000 Y+120,000
		                           RR F40 M
		24 L X+50,000 Y+160,000
		                           RR F40 M
		25 CC X+50,000 Y+130,000
		26 C X+32,000 Y+106,000
		                    DR+ RR F40 M
		27 CC X+20,000 Y+90,000
		28 C X+20,000 Y+70,000
		                    DR− RR F40 M
		29 L X+20,000 Y+60,000
		                           RR F40 M
		30 RND R+5,000
		31 L X−12,000 Y+60,000
		                           R0 F40 M

Traverse to tool-change position	TOOL CALL
	L
	
	L
	
	STOP
		32 TOOL CALL 0 Z S 0,000
		33 L Z+50,000
		                           R0 F9999 M
		34 L X−40,000 Y−40,000
		                           R0 F9999 M05
		35 STOP
		                             M

13 Programming techniques for CNC

Introduction

For many applications one of the distancing factors of early numerical control was the complexity and cost of its programming. Early NC used new programming languages (such as APT, Automatically Programmed Tools) and required computers that were then prohibitively expensive for all but the largest corporations. Since then, however, the programming of CNC tools has become a far easier and less specialist task. In many instances the emergence of CAD/CAM techniques has virtually eliminated the need to learn specific programming languages in detail.

A CNC part program contains a series of instructions that completely describe the entire machining operation. This is fundamentally different from a description of the object geometry alone, and must also specify information regarding changes in tooling and variations in cutting speed, for example.

Although CNC machine tools are still programmed manually, in an increasing number of applications part programs are generated automatically by dedicated CAD/CAM software packages that compile NC code from CAD data. This method is rapidly becoming the norm for small and large-scale applications alike, enabling almost instantaneous code production for even the most complex of forms.

Programming methods

Manual part programming

Manual part programming traditionally entailed writing the program directly in a word addressed format. This is

Part program example: machining a contoured shape. Reproduced from Heidenhain Machine Operators' manual.

normally undertaken at the machine tool, by an operator who is familiar with both the relevant machining process and NC machine code. Using engineering drawings, and perhaps a prototype, the operator must calculate the dimensional relationship between key points on the object. These points are then used to describe co-ordinates for the tool's path. The part program must specify these co-ordinates and also the sequence of machining operations, including other variables such as tool changes, spindle speed and rate of feed. As a result, the practical experience of the operator in setting these parameters cannot be overstated. Programming CNC machines in this manner can be both time consuming and tedious, particular if a part is complex, and is normally only used for simple parts and in workshops where the machine is not part of an overall manufacturing system. Despite this, its use is still widespread, as it requires no further hardware than that which is already installed within the machine.

Many CNC machines have extended manual programming into 'conversational' programming techniques that reduce the need to write each program block in its entirety. Operators can call upon existing subroutines, or macros, to describe standard features such as threads or tapers, pasting them into a larger program as required. Simulation of the machining operation is also possible and 'dry runs' may be executed with the proposed tool path displayed visually to ensure that it will not collide with other parts of the machine or work-piece. These simulations are a critical part of the process, as many 'open-loop' machines without feedback cannot detect if a collision is imminent (the large range of tooling and work-holding devices which may be fitted to many machines make automatic collision detection a complex undertaking).

Graphical techniques

Computer-aided part programming
Computer-aided part programming makes use of specific programming languages that are used to describe the key points and surfaces of a part. It differs from part programming language in that it requires a description of the object rather than the path of the tool. Conversion into NC part programming language is then carried out automatically. In practice this method has been increasingly supplanted by CAD/CAM applications that produce NC code directly from CAD data.

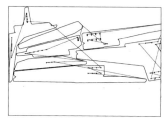

13.1
CAD generated cutting path for CNC plasma cutter. Credit: author.

CAD/CAM programming

Current CAD/CAM software packages have almost completely eliminated the need to work with part programming languages directly, and are becoming the primary means by which many CNC manufacturing processes are driven.

In this method, a CAD file containing two or three-dimensional data forms the basis of the subsequent program, which is compiled automatically by a separate dedicated software package. Compatibility to most CAD packages is generally high as data is exported in standard file formats such as .dxf or IGES. The software then converts the dimensional information contained within the file into a series of tool paths that comprise the entire machining operation.

Although CAD/CAM programming methods are commonly used for two-dimensional cutting operations, they are especially valuable in realising complex three-dimensional surfaces whose programming by manual means is often impractical. Further features provide visual simulations of the generated tool paths, and the optimisation of the machining process and its many parameters.

Digitising and reverse engineering

The ease by which object geometry can be manipulated using CAD packages has led many designers to consider its use not in generating designs in the first instance, but rather to edit and manipulate design data captured from model and prototype studies using three-dimensional scanning techniques. This allows designers to create models using traditional tools and techniques, and perhaps even has the potential to overcome much of the dissatisfaction many three-dimensional designers currently express with the standard mouse and keyboard interface.

Reverse engineering has created an important role for the real prototype in the midst of a proliferation of virtual encounters, enabling existing 'haptic' design methods to be incorporated into emerging CAD/CAM strategies. This is of particular importance to designs that are not necessarily driven by concerns of optimisation or efficiency, but are generated from a wider range of formal and spatial issues.

CNC operation

A number of control strategies are used with CNC systems. One of the primary tasks of each is to move the tool along

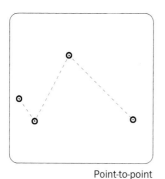

Point-to-point

13.2

the desired path as accurately as possible. Each varies in their versatility and is selected carefully to ensure their capabilities are sufficient for a required application. They differ in their capabilities of tool control and in the mathematical means by which the tool path is generated.

Point-to-point mode

Point-to-point control systems move the tool between a series of predetermined co-ordinates and machining takes place at these points only. After each operation the tool is removed from the work-piece and traverses rapidly to the next point. This simple system is well suited to the drilling or punching of holes in orthogonal work but is not capable of machining three-dimensional forms.

Contouring and interpolation

Contouring systems enable the control of the tool along a continuous path and are necessary for the machining of three-dimensional and 'free-form' surfaces. This is achieved by simultaneously varying the velocity of movement in each individual axis of travel as the tool cuts continuously along the resulting path. This capability is prevalent on the majority of CNC lathes and milling machines, as well as a growing range of cutting, welding and grinding machines.

As the surface to be machined is described in the part program as a series of discrete points, the machine's processor must *interpolate* between these points to create a continuous machining path, i.e. given that we know the position of two points we may then calculate any position between them. This interpolation can be achieved by a number of means and many machines are capable of operating in more than one mode. The following are the most prevalent.

Linear interpolation

Linear interpolation constructs straight lines between the programmed points. This path can then be machined using simultaneous movements of all machine axes (at differing feed rates). This mode is ideally suited to straight lines at any angle, but can also be used successfully for contouring. In these applications, however, curves must be approximated into a series of straight-line segments. How successfully this approximation resembles the ideal is dependent on the number of increments which forms the path. In practice this method produces good results when

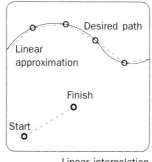

Linear interpolation

13.3

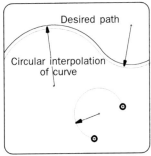

Circular interpolation

13.4

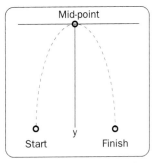

Parabolic interpolation

13.5

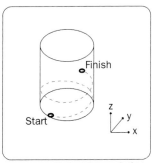

Helical interpolation

13.6

machining complex profiles, but generates large data files and requires considerable processing power.

Circular and parabolic interpolation

Circular interpolation uses circular movements between known points to construct a continuous path. This greatly reduces the number of data points to be calculated, as the mathematics of the circle are consistent for each operation. The controller is able to construct a circular path between two points given only the position of those points, the midpoint of the circle and its radius. Using this method, combinations of curves and holes may be machined without approximating to straight-line increments; however, circular interpolation can only operate on a two-axes plane at any one time. Parabolic interpolation works in a similar fashion, constructing a path between three non-straight line positions using the mathematics of the parabola. This method is ideally suited to sculpting applications.

Helical interpolation

Helical interpolation is a combination of circular and linear methods. Circular interpolation is used on a two-axes plane whilst the third axis incorporates linear interpolation simultaneously. This produces a helical path between the known points. This has obvious applications in the cutting of threads, but is also used to generate complex surfacing paths.

Feedback

Feedback enables the actual position of the tool to be compared with its calculated position. Systems with feedback, known as *closed loop systems,* monitor the position of the worktable or tool using an integral measurement system. In open loop systems the tool is assumed to be in its calculated position and accuracy must be checked by external means.

An extension of the feedback concept is *adaptive control.* This utilises a range of signals from the machining process to cope with its many variables; these may include tool wear, changes in work-piece dimensions or temperature. Adaptive control uses this information to change the operating parameters of the process automatically and in real time. Despite the advantages of feedback systems their cost still results in many machines being operated in an open loop condition, particularly in small-scale enterprises where stand-alone machine tools are used and manual supervision is still possible.

Hardinge® CONQUEST® T42 CNC slantbed lathe with two-axis top turret and single-axis bottom ram turret for heavy duty end-working operations. The turrets are programmed to operate simultaneously for minimal cycle time. The articulated pipes spray coolant onto the work-piece during machining. Credit: Hardinge Machine Tools Ltd

14 CNC lathes and turning centres

14.1
Hardinge® CNC® slantbed lathe.
Credit: Hardinge Machine Tools
Ltd

14.2
The boy mechanic's lathe.
Reproduced from James Lukin
(1885). *Turning for Amateurs*.
London: Upcott Gill.

Introduction

Despite the complex genealogy of machine tools over the last two centuries, even the most advanced of machine tools still display a lineage that begins with two of the oldest machine tool types: the lathe and the milling machine. One could simplify this still further if we consider that many of the operations of the milling machine were first carried out on adapted lathes. What is intriguing about many machine tools is their role in manufacturing their own evolutionary successors. The nineteenth-century workshops of Clement and Whitworth were able to create new and improved tools only by using those already at their disposal. This development would also be highly dependent on extraordinary manual skill, required to form components that no machine could then realise.

Even with the accelerated development of computer-aided manufacturing, consideration of these two 'primitives' still offers a relevant introduction to the capabilities of present machine tools and processes.

The lathe

Lathes are one of the simplest and oldest forms of machine tools. Simple pole lathes for the turning of wood have existed for many centuries. These were of timber construction and were often constructed within the forest itself, where raw material both for the construction of the lathe and its products was in abundance. Other early lathes were those used to bore cannon that existed from as early as the sixteenth century.

Description
Lathes are used to produce a wide range of parts that are often circular in cross-section. The work-piece, which may

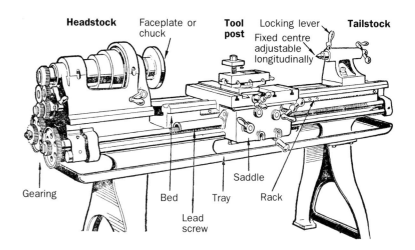

14.3
Manual lathe. Reproduced from *Odham's Practical and Technical Encyclopaedia*. London: Odhams Press, 1947.

be a length of stock material or the product of other processes such as casting or forging, is continuously rotated and brought into contact with a variety of cutting tools. Material is subsequently removed in a subtractive manner to produce the required part. The rotary action of the lathe applies an obvious constraint to the geometry of parts that it can produce, but despite this there is immense variety in its capabilities. Note that although the rotating action will always result in a circular cross-section where machining takes place, the work-piece itself can have a quite dissimilar geometry; for example, forming a tapered end to a bar of square cross-section.

On a conventional lathe each operation is controlled manually and as a result their operation requires familiarity and dexterity. As mechanisation advanced, numerous modifications were made to lathes in an attempt to reduce this dependence on skill and improve efficiency. *Turret lathes* accommodated a range of tools, each of which was often specific to a single operation and product; by using each tool in a specified sequence complex parts could be made in a repeatable manner (Fig. 3.7). Copy lathes were also developed for this purpose, which used a master pattern to control the dimensional limits of the machining operation.

Further developments in mechanisation were eclipsed by the emergence of numerically controlled machine tools. This transformed the conventional lathe into a fully automated machine tool able to machine complex forms quickly and accurately. Despite the apparent redundancy of manual dexterity, programming CNC lathes is still aided by a knowledge of machining processes, many of which are equally applicable to both traditional and emerging manufacturing techniques.

Lathe operations

Numerous different machining operations can be undertaken on the lathe. When combined they offer great scope for the production of a wide range of components. Cutting tools exhibit various geometry and dimensions and are often specific to individual lathe operations.

Turning

Turning is the most common lathe operation, in which a cutting tool is brought into contact with the side of a rotating work-piece, often to reduce its diameter to a specified dimension. Turning can produce straight, cylindrical, conical and curved parts. Grooves may also be cut into the surface with the appropriate tool.

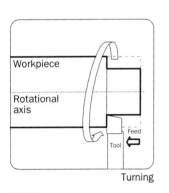

Turning

14.4

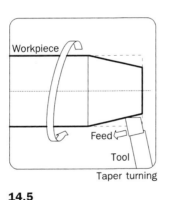

Taper turning

14.5

Facing

14.6

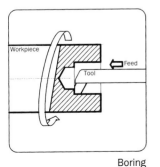

Boring

14.7

Facing

Facing involves machining the end or face of a part to produce a flat surface. This may be undertaken simply for the sake of surface quality, or where two parts will subsequently join.

Drilling and boring

Drilling is the formation of a hole within the work-piece. The drill bit, which remains stationary, is then brought into contact with the rotating work-piece. Boring is used to enlarge an existing hole, or to machine features such as grooves or internal tapers, inside the part. Boring can be undertaken to a higher accuracy than drilling alone.

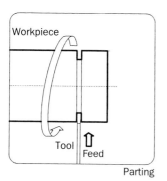

14.8

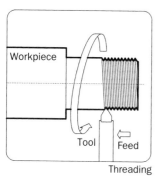

14.9

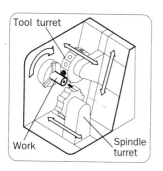

14.10
CNC lathe schematic.

Parting off

Parting off is the action of cutting a part from the remaining work-piece, normally by moving a tool into the work-piece at 90 degrees to the axis of rotation. This is often the final machining operation in the production of an individual part.

Threading

On conventional lathes, the lead-screw allows the movement of the cutting tool to be driven by the lathe motor. A system of adjustable gearing allows this movement to be synchronised with the rotation of the work-piece ensuring the cutting tool moves through a predetermined distance for each rotation of the spindle. This feature allows the cutting of threads when combined with the appropriate tool. Threads may be external (on the outside of the part), or internal (within an existing hole). On CNC lathes this operation is brought about by the independent computer control of each movement rather than by mechanised means.

Work-holding

The work-piece can be held by a number of means. A chuck, a form of rotating vice, is connected directly to the rotating spindle of the lathe. Common types have three or four jaws, and are used to hold work-pieces that are round or rectangular in cross-section respectively.

CNC lathe machine configuration

The computer control of lathes has suggested various different machine layouts, many of which differ in their operation from conventional lathes. As the capabilities of machines have expanded, CNC lathes exhibiting additional machining features are often referred to as *turning centres.*

On the majority of machines the tool-turret is no longer connected to the lathe bed as was traditional, but is mounted vertically to allow a greater range of movement. The turret houses a range of tools that are brought into contact with the work-piece for different machining operations, and is manoeuvred around the work-piece under control of the part-program. Many machines have more than one turret, each one capable of performing operations on different surfaces of the work-piece, simultaneously if required. With each of its movements under computer control the traditional reliance on the lathe bed to align each of the lathe components is removed to a large extent, as the relative position of each component and tool is derived entirely from part program control. The work is held primarily through the use of chucks, whose jaws are power operated. A tailstock is also fitted, being used to support long work, as are further devices to stabilise slender parts during machining. The most sophisticated machines incorporate 'live' tooling, such as power-driven drills, in the turret. This enables drilling, and even milling, of the workpiece, which previously would have required a separate machine tool.

14.11
Hardinge® Conquest® GT gang tooled CNC lathe. The gang tool configuration is especially suited to intricate components in large batches. Credit: Hardinge Machine Tools Ltd

14.12
Hardinge® Conquest® T 51 slantbed CNC lathe with 12-station tool turret and tailstock. Credit: Hardinge Machine Tools Ltd

Hardinge® Conquest® VNC 700
machining centre. Credit:
Hardinge Machine Tools Ltd

15 CNC machining centres

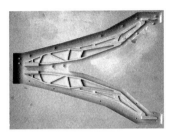

a

b

c

15.1a, b, c
Model moulds produced using a 'desktop' machining centre. Images and design: Nat Chard.

Introduction

CNC machining centres can undertake a wide range of machining operations on the different surfaces of a three-dimensional work-piece. Machining normally takes place along a minimum of three linear axes, making them ideally suited to the manufacture of non-symmetrical three-dimensional components. A further benefit of the machining centre is its ability to undertake a series of manufacturing operations that previously would have required a number of separate machines, thereby allowing individual parts to exhibit a wide range of features including contours, pockets and holes.

Machining centres can be utilised as stand alone tools, but are often incorporated within manufacturing cells as part of flexible manufacturing systems. In these high volume applications the movement of parts between machines is by means of automated pallets or modules, and the flow of materials and part program data to the machining centre is synchronised by a centralised computing facility.

The majority of machines are equipped with an automatic tool changer, enabling the use of different tools for specific machining operations. Although becoming increasingly complex, with movements in five or more axes, CNC machining centres are in many respects comparable to traditional milling machines.

Background

Together with the lathe, the milling machine remains one of the most versatile tools available to manufacturing, and the range of its applications has increased continually since its emergence in the latter half of the nineteenth

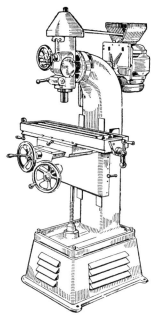

15.2
Manual vertical milling machine.
Reproduced from *General
Engineering Workshop Practice*.
This extract by E. Simons and
H.C. Town. London: Odhams,
1959 reprint. Credit: Odhams
Press Ltd.

century. Its ability to perform cutting operations in three, or more, independent axes made it the logical choice as the first experimental numerically controlled machine tool in the 1950s. Further developments in computer control have expanded considerably the capabilities of these early machines, particularly in machining three-dimensional curved surfaces. As the versatility of these machines grew, to the point where they could undertake virtually all the machining for many components, they became more commonly known as CNC machining centres.

Operation

Milling
The focus of the milling operation is the milling cutter itself. Despite vast numbers of variants, the cutter is characterised by having a number of teeth that remove the material as 'chips'. The work-piece is securely mounted on a worktable, whose controlled movement brings it into contact with the revolving cutter, thereby removing surplus material in a subtractive manner. Traditionally, milling machines were of two basic types, and the influence of these is still evident in modern CNC machines.

In *horizontal* milling machines the cutter is mounted on a horizontal spindle cantilevered over the worktable from the machine base, whilst in *vertical* milling machines the spindle is normally perpendicular to the worktable.

Although horizontal milling could often remove material faster, particularly when machining large flat areas, it proved less versatile than the vertical configuration in the geometry of the parts it could create. The differing benefits of the two configurations were often combined in a single 'universal' machine that could be built in both formats.

CNC machining centres are also found in these two traditional formats, but the computer control of linear and rotary axes now ensures that both are comparable in their versatility. Despite this, the vertical configuration remains the most common, particularly on smaller machines, whilst the horizontal configuration is often used in manufacturing cells where the worktable is palletised for transfer between machines.

The applications of these machines are virtually limitless as they are suited to a vast range of part sizes and geometry. The largest machines, which are often in a gantry

15.3
Milling a flat surface.

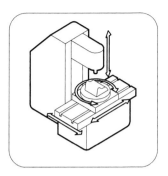

15.5
Schematic: 4-axis vertical
machining centre.

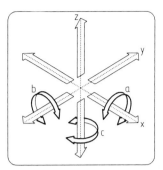

15.7
Linear and rotary axes of
movement.

15.4
Surfacing using a ball-nosed
cutter.

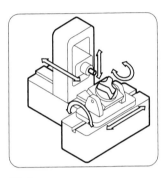

15.6
Schematic: 5-axis horizontal
machining centre.

15.8
Hardinge VMC 700 CNC
machining centre. Credit:
Hardinge Machine Tools Ltd

configuration, are capable of handling work-pieces several metres in each axis, whilst at the other end of the scale small desktop machines are increasingly available. Accuracy is typically in the region of ± 0.0025 mm.

The majority of machining centres allow movement of the work-piece in the Cartesian x, y and z-axes, each being perpendicular to each other. Machines with this three-axis movement alone are often known as *plain* milling machines, but many machines supplement these movements with an additional swivel rotation around each of the linear axes, as illustrated.

16 Case study: computer-integrated manufacturing of space grid construction systems

Process: CIM and CNC machining centres

Introduction

Space grid structures have been utilised since the beginning of the last century as a lightweight and often pre-fabricated solution for many building enclosures. However, the range of their applications has increased dramatically with the ability to model three-dimensional structures using the computer, enabling the behaviour of large and complex structures to be reliably predicted.

One of the first designers to explore space grid structures was inventor Alexander Graham Bell whose experiments, initially with kites in 1907, led to experimental tetrahedral structures for aeroplane design and pre-fabricated building construction. The scale and sophistication of space grid structures has increased steadily since that time, and a number of proprietary construction systems have been developed utilising standard joints and connections. Recent developments in computer-integrated manufacturing (CIM) have now extended the use of computer models to control the manufacturing process of these structural components, allowing the automated fabrication of both standardised and unique space grid structures.

The MERO® system

German engineer Max Mengeringhausen (1903–1990) developed the MERO system in the early 1940s as a fully prefabricated and mass-produced construction system for space grid structures. The original system used standardised steel tubes that were connected using spherical steel joints, each of which had 18 symmetrical threaded holes allowing connecting tubes to be joined at 90°, 60° or 45° to each other. The system proved a universal and highly versatile means of construction suitable for both small and large-scale space structures of many types, including solutions for skeletal grids, barrel vaults and geodesic domes. Since that time, MERO have developed additional systems utilising disc and cylinder connecting nodes specific to single layer structures, and associated enclosure systems utilising glazing and cladding.

Although the universal node system supports a diverse range of unique structural possibilities, much of the visual and spatial character of each solution was prescribed by the pre-determined angles of connection. However, many designers sought structures exhibiting varying degrees of double curvature that ideally required differing connection angles for each node and, in many instances, connection to more than one type of connecting element or to an associated enclosure system. Situations such as these demanded the production of customised components unique to a single junction.

16.1
Node fabrication: CNC machining of a spherical node. Credit: MERO System GmbH & Co.

These complex demands necessitated a flexible manufacturing strategy that can support equally the production of standardised and customised components. By using CNC machining techniques as part of a CIM environment, MERO are now able to manufacture both connectors and nodes as customised originals specific to each design. This allows the inclusion of additional connection features, but also enables the connection geometry to be unique in each case. The use of CIM in this application combines the benefits of an established construction system whilst simultaneously reducing the previous reliance on standard-isation from which the system was originally devised.

At the heart of the manufacturing process lays a computer model of the structure, which is used indirectly to provide the numerical data for the machining of each unique part. Each node is manufactured from forged steel blanks that are then further machined with the necessary holes and

16.2
Unitised packing of prepared connecting elements.
Credit: MERO System GmbH & Co.

16.4
Stockholm Globe Arena under construction. Credit:
MERO System GmbH & Co.

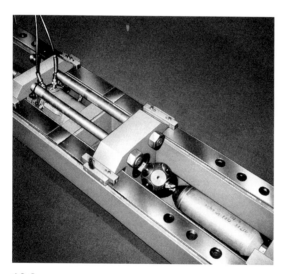

16.3
Quality control: Tension testing a node connector.
Credit: MERO System GmbH & Co.

16.5
Stockholm Globe Arena under construction. Credit:
MERO System GmbH & Co.

features unique to that connection. Information regarding connection angles and type of connector are fed directly to the machining centre from design data in a pre-determined sequence, allowing the automated machining of many thousands of unique components with a productivity that may be comparable to a fully standardised system. Once completed, nodes and connectors are also packed in sequence to enable easy erection of the structure on site.

16.6
The Portuguese Pavilion at the Hannover Expo 2000. Architects: Álvaro Siza and Souto Moura. Roof detail.

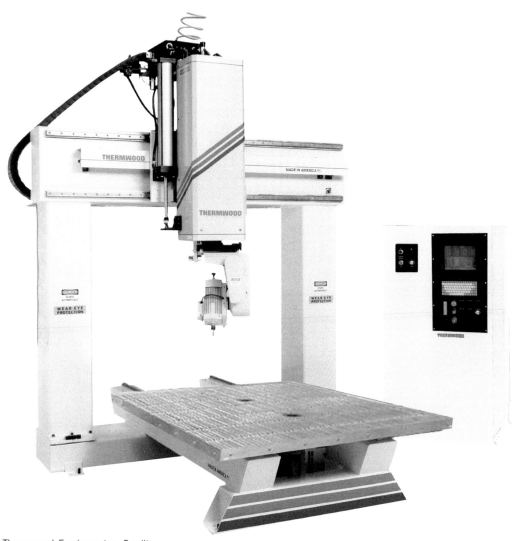

Thermwood 5-axis router. Credit:
Thermwood Corporation.

17 CNC routing

Introduction

The routing process is often compared to milling, as a rotating cutter is also brought into contact with the workpiece to remove material in a subtractive manner. However, routing differs from milling in the materials for which it is suitable and its resulting applications.

The first CNC applications were almost exclusively within the metalworking industries, and these early machines were consequently designed for relatively slow rates of machining but with high levels of precision. As CNC technology became more widely disseminated further applications soon became apparent. Within the woodworking and furniture industries the manual routing process had been used for decades to rapidly machine rebates and grooves in joinery, and their evolution into automated techniques suggested not only benefits in productivity but completely new methods of manufacture. Similarly, in the plastics industry various manual and mechanised methods had

17.1
Thermwood C42 3-axis CNC router with turret tooling. Credit: Thermwood Corporation.

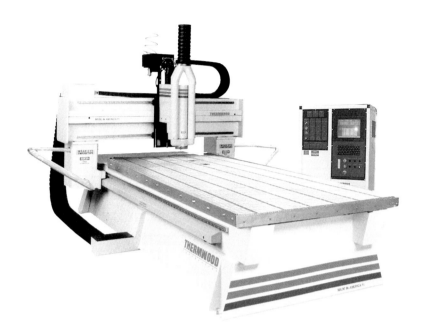

17.2
Thermwood C53 3-axis CNC router. Credit: Thermwood Corporation.

been used to trim and finish plastic mouldings, but most were considerably less efficient than the moulding process itself. CNC routing offered opportunities to both these domains and suggested still further applications in sign manufacture and mould making.

The routing process is still generally used for wood and plastic products rather than metals, and material is consequently removed at much higher rates. Router cutters will pass through the material at 10 to 100 times faster than in the milling of metal, and typically rotate at from 18 000 rpm to as high as 60 000 rpm. The dynamics of this process have generated unique machine types that differ increasingly from traditional metalworking precedents.

Machine configuration

Routing is effectively a group of processes that encompass drilling, slotting, grooving and shaping operations, amongst others. Machine design takes many forms, varying in scale from those suitable for model making, which fit on a table-top, to large machines capable of finishing rotational plastic mouldings up to 1.5 m in each axis and beyond. Three and five-axis machines are the most common, and whilst three-axis control theoretically allows the cutter to reach any point within the work envelope this is often not possible in practice as the work-piece itself becomes an obstruction. This is overcome in five-axis machines by providing the cutting head with two rotational axes perpendicular to each other, providing greater access to overhangs or internal features, for example (Fig. 15.7).

Many routing applications use panel products such as MDF and plywood as their stock material so the typical work envelope for a machine is normally greater in the x and y-axis than in the z-axis. The most common machine configuration is a 'gantry' type, with either a fixed or moving table on which the work-piece is mounted. Moving table types are often more economical, and sometimes more accurate, than their fixed counterparts but pose limitations on work-piece size and weight (Fig. 17.1). These are overcome in fixed table designs through the use of a moving gantry, but its design is often complex as it must be rigid enough to resist forces generated during cutting yet still be able to move swiftly and accurately (Fig. 17.2). Commercial CNC routers can currently machine at up to 30 m per minute and reproduce parts to within 0.05 mm in the best cases.

Machining operations

The form and size of the cutting tool itself distinguishes the various machining operations possible with routers. As with the majority of CNC machines, tool changing is often an automatic procedure carried out under part program control. First generation three-axis machines provided general routing processes including drilling and edge profiling, but current five-axis configurations have extended this range to include sawing, contouring and carving.

Routing
The primary routing process utilises a fluted cutter of relatively small diameter, less than 25 mm, say, which rotates at high speed. Routing is used to cut smaller shapes from panels and to create further features such as rebates and grooves. The cutters themselves may exhibit a straight or profiled geometry depending on the desired cross-section.

Shaping
Shaping is used to generate specific profiles on internal or external edges, often with a higher finish and rate of material removal than routing. Many shaping tools incorporate interchangeable blades that fit within a universal head, and

17.3
Routing tool. Credit: Thermwood Corporation.

17.4
Shaping tool. Credit: Thermwood Corporation.

17.5
Horizontal drill. Credit:
Thermwood Corporation.

17.6
Circular saw. Credit: Thermwood
Corporation.

17.7
Sanding head. Credit:
Thermwood Corporation.

overall cutter diameter typically reaches 150 mm and beyond. Moulding is a related process, and is normally thought of as shaping in a straight line. Complex profiles can be achieved by using a number of different shaping tools, each removing material in successive operations.

Drilling and boring

Holes can be machined by several methods. For smaller diameter holes, an appropriately dimensioned router cutter is plunged into the work-piece in a single operation. For larger diameter holes the outline perimeter of the required circle is routed and the remaining disc discarded. Dedicated drilling heads are also available, which are often necessary to drill panel edges and other surfaces that are difficult to access (Fig. 17.5).

Sawing

Whilst routers are well suited to machining both linear and free-form features, in high volume applications linear cuts are more efficiently undertaken using a rotating circular saw blade. Some manufacturers now produce tooling attachments specifically for this purpose to compliment other routing operations (Fig. 17.6).

Sanding

In the woodworking industry in particular, the sanding and finishing of parts takes up significant portions of the overall production cycle. Sanding heads are similar in their construction to shaping heads, and are often used to impart a final finish to previously profiled edges (Fig. 17.7).

Squaring

The action of the rotating cutter in the majority of routing operations dictates that internal openings and features can only be machined with a radius, which corresponds to that of the cutter itself. Where this is not desired squaring tools have been developed to remove this region of excess material. They typically utilise a number of scraper blades that oscillate back and forth towards the corner, removing material in a number of stages to achieve a good finish.

Carving

The use of CNC routers for carving was first introduced in the early 1990s, and draws upon long established uses of CNC machining centres in the metalworking industry to produce contoured three-dimensional surfaces.

A number of different cutter types may be used in any single carving application. Initially, the bulk of the excess material is removed using a ball-nose cutter, whose spherical form can readily accommodate the contoured surface. After the shape has been roughed out, flat surfaces are levelled using square-ended cutters before fine details are created using a pointed detail tool. CNC carving and its associated tool path generation require fast processing speeds and create large part program files.

Source: Susnjara, K.J., (1999) *Three Dimensional Trimming and Machining: The Five Axis CNC Router*. Dale: Thermwood Corporation.

A worker programming a CNC
plasma cutting machine at Ehlert
GmbH, Güsten. This model can
also utilise the oxy-fuel process.

18 CNC thermal cutting systems

18.1
Plasma cutting nozzle.

18.2
Plasma cutting. Credit: Messer
Cutting & Welding AG.

Introduction

Most materials, including metals, will burn. A number of cutting processes exploit this fact by bringing about the rapid oxidation of material, to form a kerf, or cut. Thermal cutting in this manner offers many advantages over traditional subtractive machining methods such as sawing. As the process does not require the physical contact of a tool with the work surface, cutting forces are almost negligible, rates of material removal are faster and the mechanics of machine design are simplified.

The cutting of steel and non-ferrous metals using CNC plasma and CNC laser cutting techniques is well established, and is used by large and small-scale enterprises alike. Developments in laser cutting have also increased the range of materials suitable for thermal cutting to include non-metallic substances such as wood, plastics and ceramics.

Background

The cutting of metals using rapid oxidation is perhaps best known through the use of oxy-fuel gas cutting. This familiar process, which is still widely used, can be recognised in the flame-cutting torch and gas bottles used by demolition firms, welders and mechanics alike. The equipment is portable, requires no electrical power and, using the appropriate accessories, can also be used for heating and welding.

In the oxy-fuel process a fuel gas, often acetylene or propane, is mixed with pure oxygen within the torch (which may be hand-held or part of a larger machine). This quickly raises the temperature of steel to 900°C, when it will glow orange-red. A jet of high purity oxygen is directed at the heated area causing the steel to rapidly oxidise. A continuous cut is formed as the torch, maintaining the heat and oxygen, is traversed across the surface of the work-piece.

Until the development of plasma and laser cutting techniques, oxy-fuel cutting was the most popular means to cut shapes from steel plate and sheet, becoming an essential technique in the fabrication of steel structures of many kinds. Before the advent of CNC, tracer machines were often used, which followed separate templates to cut a variety of shapes. CNC plasma and laser cutting machines support similar applications, but can offer significant improvements in speed, quality and material choice. Despite these recent advances oxy-fuel processes are still widely used, particularly when thicker material must be processed. Oxy-fuel, plasma and laser cutting machines now all utilise CNC technology, and some manufacturers produce dual-purpose machines that support both oxy-fuel and plasma processes. Most machines are constructed in a modular form and large work-piece envelopes are possible; in practice work-piece size is more likely to be constrained by transport considerations.

Plasma cutting

Plasma arc cutting was originally developed to cut non-ferrous metals, for which regular oxy-fuel processes are not suitable. It is now commonly used to cut both ferrous and non-ferrous metals in both CNC and manually operated formats.

Plasma has been described as the fourth state of matter, and consists of an ionised gas at temperatures as high as 9000°C. In the cutting process plasma is generated as an inert gas (or air) is combined with the arc from an electrode. The gas becomes a plasma as it is channelled past the electrode, forming an arc stream of high temperature ionised gas. The resulting cut is narrow with a surface finish that requires little further finishing for most applications. The process is ideal for thin and medium thickness mild steels, and is also effective for stainless steels of up to 150 mm in thickness.

Plasma marking
The plasma process can also be used to mark bend lines and other graphic features through the deposition of zinc powder upon the work surface.

a

b

c

d

18.3a, b, c, d
Typical plasma cut parts. Credit:
Messer Cutting & Welding AG.

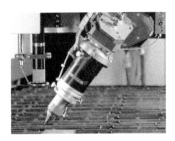

18.4
Laser rotation head for exact bevelling or the exact cut of inner contours. Credit: Messer Cutting & Welding AG.

Laser cutting and machining

The laser is a powerful and controllable source of thermal energy that has found many applications in cutting, machining and welding. Several different types of laser are used, each favouring different applications, although the CO_2 and Nd.YAG (neodymium:yttrium-aluminium-garnet) types are perhaps the two most commonly used.

The laser
The laser (an acronym for Light Amplification by Stimulated Emission of Radiation) exploits the fact that electromagnetic radiation is emitted when the electrons surrounding an atom's nucleus change energy states.

The electrons surrounding any atom nucleus always assume discrete orbital positions corresponding to their energy level. Collisions with other atoms transfer energy to these electrons causing their movement to a higher level. Without further energy input the electron will lose this 'excited' state and return to its original state, releasing surplus energy as a photon. The frequency of this electromagnetic radiation is specific to each atomic substance.

In industrial lasers this photon emission is induced by creating a collision between an excited atom and an additional photon. This creates two photons 'in phase' that initiate a further chain reaction of amplified emission. The phenomena gains practical effect when the resulting electromagnetic wave is amplified still further in a resonator, creating a parallel light beam of one wavelength with a power of up to 6000 kW. Additional optics are incorporated to direct and focus the beam on to the work surface.

a

b

18.5a, b
Typical laser cut parts. Note the sharp corners and close spacing of features, 5 mm mild and stainless steel. Credit: Messer Cutting & Welding AG.

Laser cutting operations

Laser sublimation cutting
Used mainly for non-metallic substances such as wood, paper, ceramics or plastic, laser sublimation vaporises material at the cutting point. It requires a high intensity beam and normally takes place within the atmosphere of an inert gas. This is pumped onto the working area to prevent oxidation from atmospheric contact.

Laser fusion cutting
Also known as high-pressure cutting, the material to be cut is first melted by the beam and then cleared from the kerf

with nitrogen gas at high pressure. The process is particularly suitable for stainless steels and aluminium. The lower temperatures of the process and the nitrogen shielding gas seek to ensure a cut free of oxide contamination.

Laser burning

In this variant the laser beam heats the material to its ignition temperature before a jet of oxygen is directed at the working point. The material burns in the oxygen-rich atmosphere forming the kerf. The kinetic energy of the gas stream also clears the oxide slag from the kerf as it is formed. This process is often used for mild steel.

Laser marking and machining

The power output of lasers is highly controllable. Low power applications have been found in the marking of surfaces for labelling and setting out purposes. Similarly, lasers are finding applications in specialist drilling applications where conventional 'contact' methods are costly or impractical. This is often the case with extremely hard materials or when forming small diameter holes at oblique angles to a work surface.

Process comparison

Laser

Suitable for

- Thin and medium thickness materials (up to 20 mm).
- High accuracy.
- Square edged cuts with most materials.
- Narrow heat affected zone.
- Oxide free cuts in stainless steel.
- Non-metallic materials can be used.
- Complex component shapes possible, even at small scales.
- Part features, e.g. holes and slots, can be closely spaced.

Materials: mild steel, stainless steel, aluminium, plastic, wood and textiles.

Plasma

- Suitable for thick stainless steels (up to 150 mm).
- Suitable for thin and medium thickness mild steel.
- High cutting speeds.
- Normally more economical than laser cutting.

Materials: mild steel, stainless steel, bright alloys.

Oxy-fuel

- Suitable for medium and heavy thickness mild steel (up to 300 mm).
- Simple shapes.
- Accuracy not critical.

Materials: mild steel.

Machine configuration

The majority of CNC thermal cutting machines are of a 'gantry' type, where the cutting head traverses a beam that itself travels the length of the working area. This allows simultaneous movement in the x and y-axes and the cutting of two-dimensional shapes. Movement in the vertical z-axis is more limited, and exists simply to accommodate different material thickness and to ease setting up procedures. This is sufficiently versatile for most applications as the majority of parts are cut from flat plates and panel products. Some laser cutting machines are now fitted with rotating cutting heads that permit the cutting of bevels and other three-dimensional features.

The use of CNC cutting machines has transformed the fabrication process in many workshops. Individual plates and parts would previously have been marked manually within the workshop, and cuts were often made with the assistance of full size patterns and jigs. Manufacturers of CNC cutting equipment have developed a range of software to support their machines. These packages allow automatic NC part program production from CAD drawings imported in a variety of formats (although .dxf is most common), and also enable optimisation of the cutting process itself. Each of the process parameters, such as cutting speed and power, are described within the program and little further intervention is required at the machine. Additional features, to enable the efficient nesting of parts or to diagnose cutting defects, are also available.

Electrical discharge machining

Electrical discharge machining (EDM) uses successive spark discharges to gradually erode away excess metal. The process relies on creating a high potential difference between a specially shaped electrode and the work-piece to be machined, both of which are immersed in a non-conducting fluid (dielectric). If the two are brought into close proximity, a spark discharges through the fluid eroding the metal where it makes contact. The discharge is repeated at frequencies of up to 500 Hz as the electrode is moved through the material, normally under computer control. The process exerts no mechanical forces on the work-piece, and rates of material removal are not affected by material hardness, although the work-piece must be a metallic conductor.

The process has many applications, normally being used where conventional cutting tools would be too fragile, such as very small diameter deep holes, and when machining very hard material, such as the steel used for dies and moulds. The ability to shape the electrodes, which can be as small as 0.1 mm in diameter and normally made from graphite or brass, means complex internal features can be machined at even a small scale. Despite this ability the process may also be used on large work-pieces, such as the dies required for large sheet metal presswork.

Wire-EDM
Further applications are supported by Wire-EDM, in which the electrode is a long thin wire, of tungsten or non-ferrous metal, tensioned between two adjustable reels. This works in a manner not dissimilar to a band saw, but again without

18.6a
A complex die machined using the wire-EDM process.
Credit: Charmilles Technologies.

18.6b
A Wire-EDM machine. Credit: Charmilles Technologies.

any mechanical force on the work-piece. This process is able to cut through very thick (up to 300 mm) and hard metals, and intricate shapes are possible as the electrode is typically only 0.25 mm in diameter. Again, many wire-EDM machines utilise computer control to monitor the machining process and to control the electrode path, often in more than one axis.

18.7
Wire-EDM. Schematic diagram of a 4-axis machine.

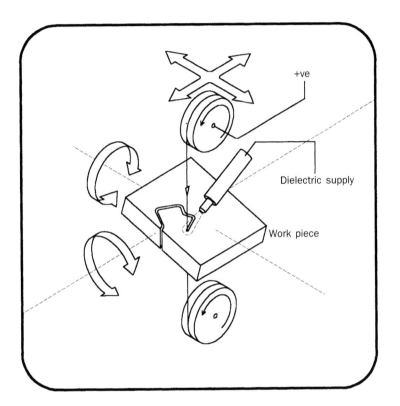

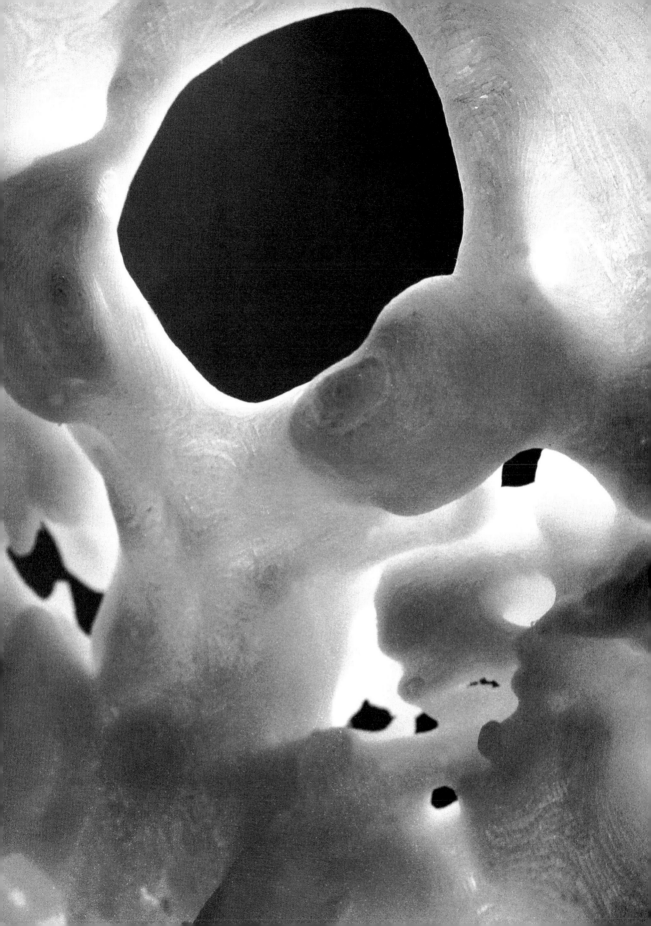

19 Solid freeform fabrication

19.1
3D Systems SLA 7000 Stereo Lithography System. Credit: 3D Systems Europe Ltd.

Prototypes

In many respects, physical prototypes exhibit a clarity and legibility that few forms of visual representation can attain. Realised by plastic techniques, the virtue of the prototype lies within the interactive modality it supports and the subsequent critical response this initiates. Its creation is at the heart of a transition from conception to execution, and the realisation of prototypes has traditionally relied on the combined resources of both designers and makers alike. This collaboration is at the root of the prototype's assured completeness as a statement of design intent, independent of its actual refinement or functionality.

Simple or complex, prototypes are invariably the product of many different cognitive and physical processes, whose fabrication is likely to utilise a host of manual techniques that differ fundamentally from those of the design's ultimate manufacture.

Detail of a skull model. Fabricated using the fused deposition modelling process at the Department of Medical-Physics, University College London.

a

b

19.2a, b
Parts produced using the selective laser sintering process on a DTM Sinterstation™. Credit: DTM Corporation.

The increasing capabilities of computer modelling and simulation have now provided us with a sophisticated visual means to interact with emerging designs. As the boundaries of what is analysable continue to expand, 'virtual prototypes' are now able to represent not only geometric data but also the physical and behavioural characteristics of complete complex systems. Simulation has allowed many of the benefits of physical prototyping to be obtained without the need to construct them in reality, and in the manufacturing industry this has significantly reduced the 'lead-time' for many products.

Despite the validity of these methods, physical prototyping has remained a key stage on the path to production. Many designers feel the use of physical prototypes within the design process remains a necessity, as many sensory and cognitive aspects of their use appear to defy simulation – indefinitely. However, prototypes are not simply valuable in initial design development and testing, but are also often necessary in the evolution of the production tooling required for many manufacturing processes. For example, die-casting, pressing and plastic injection moulding all require the production of one-off moulds specific to individual parts. As a consequence, despite the success of computer models to visualise designs, the prototyping of production tooling must by necessity take place in the real domain.

For many sectors of manufacturing, the productivity advantages of computer modelling only served to highlight existing bottlenecks in the production process, many of which were caused by the time-consuming fabrication of prototypes. For many industrial designers the apparent ease of visualisation was increasingly at odds with the more protracted process of pre-production and manufacture. Although CNC machine tools already provided a partial solution to these concerns, being an established means to fabricate objects directly from numeric CAD data, these processes were not suitable for all applications.

Industry searched for a more direct means to translate computer models into physical prototypes; one that could be used both in a non-industrial environment, and which might also overcome some of the geometric constraints of subtractive machining processes. After prolonged development, 3D Systems of California unveiled its first 'stereo-lithography' machine in 1986. This is considered the first of the 'rapid prototyping' machines, so named to reflect its ability to bypass time-consuming pre-production stages by fabricating physical models from the data within a 'solid-model' CAD file. Whilst use of the term 'rapid prototyping' is still widespread, it is now also referred to as 'solid freeform fabrication' (SFF), a term which better reflects the nature of the process rather than one specific advantage of it.

An interdisciplinary approach

Despite originating within manufacturing, other disciplines quickly saw applications within their own fields. Artists saw possibilities for new forms of sculptural exploration; medicine immediately saw applications in bone reconstruction and the development of improved prosthetics; archaeologists saw opportunities to replicate and archive fragile artefacts. The number of potential applications seemed limitless but the enormous cost of SFF equipment, particularly in the first few years of its use, meant that few single enterprises could support its capital and running costs alone.

This economic concern, together with an increasing awareness of the interdisciplinary nature of the technology, has spawned a number of research collaborations and commercial 'bureaus'. These offer a centralised SFF facility that can often be used on an hour-by-hour basis by those unable to run their own equipment.

In the long term the cost of this technology seems likely to fall to levels compatible with more established computer peripherals; some recent systems are now comparable in price to colour reprographic equipment. Yet, despite these changing economic considerations, the universal nature of these techniques will continue to generate a number of interdisciplinary projects. Future watchers have already mused as to the extreme possibilities of the technology. Home shopping, some have prophesised, will come to mean that products bought over the Internet are actually fabricated within our own homes, whilst an earlier application is possible in the use of SFF techniques to manufacture mechanical parts aboard naval warships, thus obviating the need to carry spares.

Process description

Solid freeform fabrication, also known as rapid prototyping or layered manufacture, is a generic term describing a group of processes that form objects by the repeated deposition of individual layers, or drops, of material until the part is complete. A process specific data file, derived from a CAD solid-model, controls the process.

SFF differs fundamentally from traditional machining processes in that it forms parts in an *additive*, or *constructive*, manner. Other machining processes, such as turning, milling and cutting, all form objects by the *subtractive* removal of material from the original stock.

Additive processes have a number of advantages over subtractive methods:

- *Realisation of complex geometric form.* Subtractive methods require a cutting tool to come into contact with the work-piece surface to remove material and obtain its required geometry. When forming the object in an additive manner, material is readily deposited at all points – even within the object. Deep grooves or internal voids are examples of features that are a challenge to subtractive techniques, yet are easily achieved with SFF.
- *Reduced lead times for unique parts.* Unlike many machining operations, no unique jigs, moulds or work-holding devices are required to produce the object. All preparation is in the manipulation of the virtual prototype alone, and the necessary conversion to the required .stl data file is an automatic feature on the majority of computer modelling packages.
- *Clean production environment.* Many SSF processes are completely enclosed, quiet and produce minimal waste. In many instances this even allows their installation into non-industrial environments, such as the design studio or office.
- *Allows 'non-expert' use.* SFF does not necessarily require its users to be trained manufacturing engineers or machinists, nor does it require knowledge of specialist machine tool programming languages.

Despite these advantages, there are number of occasions where subtractive methods continue to be more appropriate:

- *Material choice.* A much greater range of raw materials can be utilised with conventional machining.
- *Accuracy.* Conventional machine tools are capable of machining to far smaller tolerances with improved surface finishes.

- *Object size.* Far larger objects can be fabricated with conventional machining at this time.
- *Batch size.* For medium and large batches subtractive machining is likely to be both faster and more economical.

Applications

Visualisation, testing and manufacture

Early applications of SFF techniques were almost entirely orientated towards the production of individual prototypes for visualisation purposes. This reflected the limited palette of build materials (often polymer resins or plastics) and the dimensional constraints of early machines. Further applications were soon sought in limited physical testing, however, including wind tunnel and optical stress analysis.

More recently, developments have focused on extending the benefits of SFF into the manufacturing process itself. SFF techniques are now used to make master patterns for casting operations and other forms of production tooling including dies and electrodes. This often utilises powder metallurgy techniques, which allow the prototype to be replicated in metallic materials without machining. This field, known as rapid tooling and rapid manufacturing, now represents a significant percentage of SFF applications.

Input data file formats

SFF machines of most types require data to be input in a standard file format known as **.stl**. Originally developed by 3D Systems® to support their stereo lithography process, it has now been adopted as an industry standard for all SFF processes. As a result many CAD and modelling applications provide an **.stl** export facility for objects drawn as solid models. In specialist applications .stl files can also be generated from three-dimensional scanning and experimental data.

.stl files describe geometry by mapping the complete surface of an object as a series of adjoining triangles, and then describing the co-ordinates of the vertices. Whilst this is a somewhat inefficient means of describing geometry its use has persisted, an occasion where standardisation has won over elegance.

To create an .stl file the original solid model must be a 'legal' solid. This requires it to be modelled as a continuous solid with no areas of zero thickness. In addition the surface must not pass through itself, as this forms an overlap of solid regions. In practice, these constraints need never limit the desired geometry of a part, but simply dictate how it is drawn, requiring close attention to solid and surface conditions.

Additional software, installed on a workstation connected to the SFF machine, reads the .stl file and 'slices' the object into a series of layers, which correspond to those in the subsequent model.

SFF processes

Stereolithography
Stereolithography (SLA®) forms parts by selectively curing a liquid epoxy resin.

Objects are fabricated on a movable platform that is immersed in a resin vat. At the beginning of the build the platform is positioned just below the surface of the liquid. A laser scans the surface of the polymer with ultra-violet light, selectively curing only the areas described by the object's geometry at that layer. Having cured this layer, the platform is lowered a small increment to create a further layer of uncured resin above the first. This is then similarly exposed and cured. The process is then repeated until the object is completely fabricated, at which point the platform is raised and the object removed. Excess resin is then washed off, and the whole object further exposed to ultra-violet light to prevent contamination when handling.

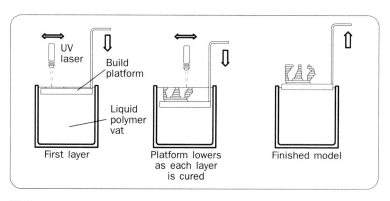

19.3
Schematic: Stereolithography process.

Specification:

Typical build envelope	500 × 500 × 500 mm
Build material	Epoxy resin
Input file format	.stl and .slc
Resolution	Minimum layer thickness typically .0025 mm
Applications	Visual models
	Optical stress analysis
	Snap fit assemblies
	Wind tunnel models
	Underwater testing
	Master patterns for further mould making and casting processes

SLA is a trademark of 3D Systems

Selective laser sintering

Selective laser sintering (SLS®) forms parts by selectively fusing together a polymer coated powder.

Parts are formed on a movable platform, which is repeatedly covered in a thin uniform layer of powder. A roller spreads and smoothes the powder surface, which is sintered (fused together) using a CO_2 laser at selective points described by the data file. The platform is lowered a small increment, and the roller spreads a further layer. The process continues as the part under construction is lowered progressively into the build chamber, supported by loose powder. Once complete the object is removed and traces of the loose powder cleaned away manually.

Specification:

Typical build envelope	380 × 330 × 450 mm
Build materials	Plastics (rigid and elastomeric), metals and ceramics
Input file format	.stl
Applications	Functional models
	Sand casting cores and moulds
	Expendable patterns for investment casting
	Mould inserts for plastic injection moulding
	Mould inserts for pressure die-casting

SLS is a trademark of the DTM corporation

Laminated object manufacturing

Laminated object manufacturing (LOM) builds objects from layers of paper or plastic coated with a heat-activated adhesive.

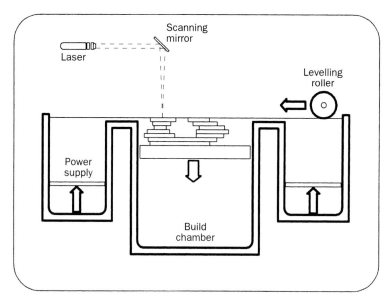

19.4a
Schematic: Selective laser sintering process.

19.4b
DTM Sinterstation®. Credit: DTM Corporation.

19.4c
Rubber-like prototypes can be produced using an elastomeric polymer as a build material. Credit: DTM Corporation.

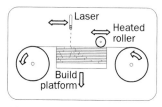

19.5
Schematic: Laminated object manufacturing process.

Objects are formed on a movable platform positioned between two spools that transport the sheet material. Material is loaded from the supply spool, passes over the fabrication platform and onto a similar take up spool. A laser cuts the outline of the object corresponding to that layer, as well as cross hatching the waste area – this allows the later removal of the excess by manual means. A heated roller passes over the area, successively bonding each layer to the next. The process is popular for large models which have the intriguing property of feeling and looking like wood.

Specification:

Build envelope	760 × 500 × 500 mm
Build material	Paper and plastic
Input file format	.stl
Layer thickness	0.005–0.05 mm
Applications	Functional models
	Wind tunnel tests
	Visualisation models

Thermojet®

3D Systems' ThermoJet is one of a growing number of solid object 'printers' that are conceived from the outset to be used within a studio or office environment in much the same way as any other computer peripheral. Utilising inkjet technology, tiny droplets of a wax-like substance are sprayed onto the build platform, forming the object layer by layer.

Specification:

Typical build envelope	250 × 190 × 200 mm
Build material	Thermopolymer
Input file format	.stl
Resolution	300 dpi
Applications	Visualisation models
	Master patterns for casting

Fused deposition modelling

Fused deposition modelling (FDM®) forms objects by the controlled extrusion of melted plastic.

Heated modelling filament is passed through an extrusion head under computer control. Material is deposited as a series of fine lines that fuse the surrounding material as each layer is built. The object is fabricated on a fixed platform.

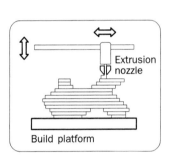

19.6
Schematic: Fused deposition modelling process.

Specification:

Typical build envelope	250 × 250 × 250 mm
Build material	ABS plastic, investment casting wax, elastomer
Input file format	.stl
Layer thickness	0.05–0.8 mm
Applications	Visualisation models
	Investment casting patterns
	Functional models

FDM is a trademark of Stratasys products

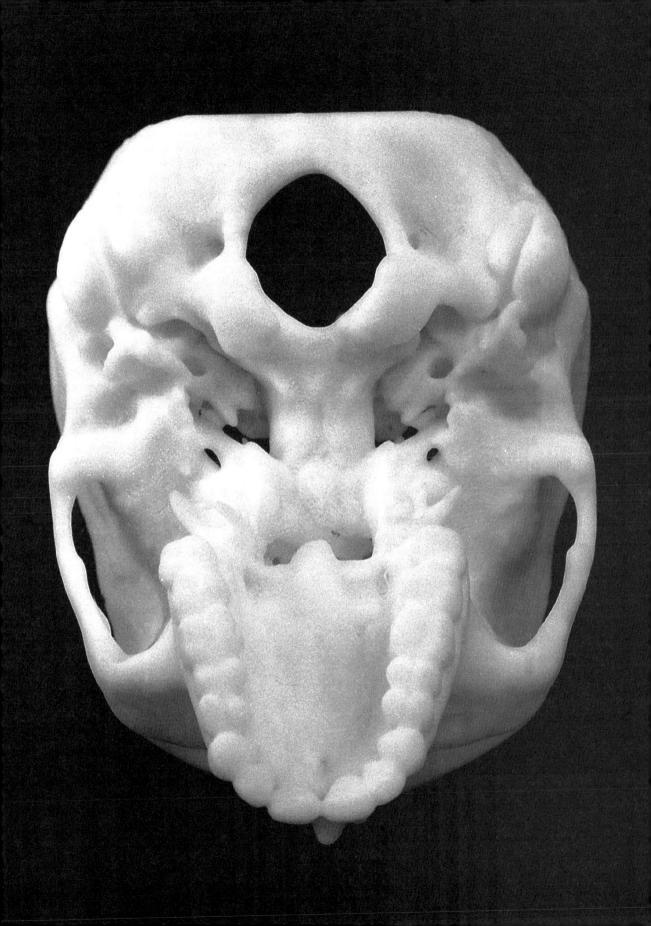

20 Case study: re-constructive surgery

Process: Solid freeform fabrication: fused deposition modelling, FDM®

Within the maxillo-facial unit at University College Hospital London, surgeons seek to remedy a variety of anatomical conditions that have arisen due to congenital defects or as the result of accident trauma.

The surgery required is invariably complex, and requires that the surgeons involved have a detailed visual and spatial knowledge both of general facial anatomy and also of the patient's unique condition. Many procedures involve the insertion of titanium implants to repair or replace defective areas of bone structure, which are customised to the patient and must be fully fabricated before surgery commences.

An important requirement of the implants is that they fit adjacent bone structure accurately; otherwise surgical procedures may have to be prolonged whilst alterations are made. Constructing the implants from two-dimensional x-rays or scan data alone is extremely difficult and technicians ideally require a full-scale model of the patient's anatomy for use within the workshop. This provides a replica against which parts may be tested for fit and shape before surgery.

The Department of Medical-Physics UCL has been experimenting with a number of CAM techniques to realise these models. In each case three-dimensional scan data is processed into a computer model of the patient's bone structure. This data is then used to automate machining of the replica. This was initially undertaken using a three-axis CNC milling machine, but because of the many intricate cavities in the skull's structure this subtractive method could not produce a true rendition of the affected region.

Full-size skull model in ABS plastic. Fabricated using the fused deposition modelling process at the Department of Medical-Physics, University College London.

20.1a
Model detail.

20.1b
Partially built model – the darker parts are support structures. Generated automatically by the machine's software they support overhanging features during fabrication. They are subsequently removed by dissolving in solvent.

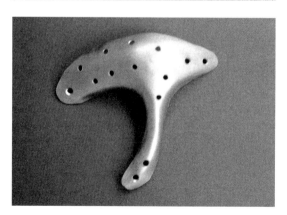

20.1c
A patient's titanium implant.

These limitations have now been overcome through the use of solid freeform fabrication techniques, whose additive fabrication allows the realisation of even the most complex internal features. The resulting models also have further benefits, as they provide surgeons with an invaluable means of preparation, enhancing the knowledge gained from scans, and providing an ideal means to describe the procedure to the patient.

In the future, 'rapid prototypes' may be used directly within the body itself, but at present issues of sterilisation and rejection prevent this. As a result the implants themselves are still crafted by manual means.

FDM is a trademark of Stratasys products

21 Production tooling

'Tools make tools'

The hardware of production does not consist of machines alone, but also includes an assortment of other devices and attachments with which they are combined during

21.1a
Product specific tooling: sculptor John Bremner designed and hand-crafted his own press-tools to fabricate his untitled aluminium piece in 1996. It consists of approximately two thousand identical parts. Credit: John Bremner.

A water-cooled mould from a continuous casting machine, producing universal beam blanks, c. 1970. Credit: Science and Society Picture Library.

21.1b
Two part stamping tool for producing aluminium blanks.

21.1c
Two part press tool for shaping the blanks.

21.1d
Finished components.

manufacture. Inherent to the manufacture of any item is the need for production tooling which is often unique to the product and its manufacturing process. Knowledge of the individual nature of the tooling required for specific processes is essential to assess the economic viability and technical feasibility of any individual design.

CAD/CAM techniques have had a large influence on the nature and fabrication of production tooling directly reducing the time, cost and physical resources required for manufacture.

Generally speaking, tooling allows the range of actions, or the variety of products, which a single machine can undertake to be extended without fundamental alterations to the machine itself. In some cases tooling is essential for the manufacture of a particular part; in others, it allows the efficiency or accuracy of a process to be improved.

Production tooling can include cutters, dies, jigs and moulds, each of which are used in conjunction with a machine or process to undertake a particular task. Perhaps the simplest example of the tooling concept is that associated with the use of the electric drill, whose adjustable chuck allows its use with a variety of different sizes and types of drill bit, each suited to different materials. The drill's range of uses can also be further extended by attachments for the driving of screws and other mechanical fasteners. In this way a single device is able to undertake a considerable range of tasks more economically than the provision of a specialist tool for each task. While this is obvious in such an everyday example, within manufacturing industry considerations regarding the nature of tooling are fundamental to the selection of appropriate methods of fabrication. It is often the deciding factor when considering whether particular manufacturing processes are viable for a specific batch size.

Process and product specific tooling

Tooling can be divided into two broad categories, being specific to either process or product.

Process specific tooling is designed to undertake a specific machining activity or process in a specified range of materials. The drill bits in the above example belong to this category. They are not unique to the part or product being made, but rather can be used towards the creation of a

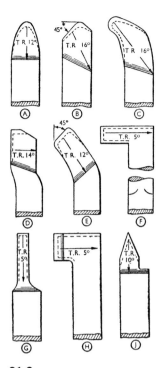

21.2
Process specific: a selection of lathe tools. Reproduced from *General Engineering Workshop*. This extract by H.C. Town. London: Odhams Press Ltd, 1959 reprint. Credit: Odhams Press Ltd.

21.3
Process specific: drills and reamers for forming holes.

range of items that each require the same feature. Cutting tools for lathes or milling machines are other universal examples (Figs 21.2 and 21.3).

Product specific tooling is unique to a single component type and their size and geometry is related directly to that of the part required. Patterns for casting and dies for the extrusion of plastic or aluminium sections are typical examples. Tooling may be necessary for the fabrication of a single item or, more commonly, to allow the rapid formation of identical parts in quantity, as in the two-part tools used to press metallic panels. Tooling of this category has often been used to form components with complex surface geometry, such as car body panels, which could only be made economically in numbers using processes that could be readily mechanised. By creating a tool with a corresponding geometry to the required part it allows production through one single mechanised action, rather than a hundred or more skilled manual ones (Fig. 21.1).

CAD/CAM and tooling evolution

The fabrication of tooling is often a considerable undertaking in itself. The universal nature of process specific tools such as the anonymous drill bit maintains their production in large numbers by fully automated means, but product specific tools are often unique and complex originals. To fabricate a curved press-tool, for example, the toolmaker may have to make the combined use of several machining processes in conjunction with highly skilled handwork. The previous inability to automate the fabrication of production tooling had been at the root of its time consuming and costly nature, which could often only be justified when destined to make large quantities of parts. Many of the implications of CAM to the production path are at their most positive in the development of tooling, and have now reduced many of its constraining factors. Consequently, designs previously too expensive to realise at smaller batch sizes may now be realistic proposals for the first time, whilst the palette of manufacturing processes applicable has also expanded.

CAM affects the realisation of production tooling in the following ways:

1. *Direct manufacture: the elimination of tooling.* CNC processes allow the automated machining of components under the direct control of data from a part

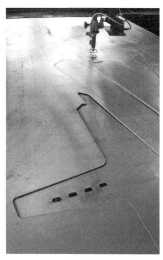

21.4
Automated cutting of steel
component using a CNC plasma
cutter.

21.5
CNC machined casting mould.
Image and design: Nat Chard.

21.6
Demonstration casting pattern
fabricated using selective laser
sintering. Credit: DTM
Corporation.

program, often compiled from CAD representations, and consequently require little or no product specific tooling for manufacture. The processes involved are often subtractive in nature, as material is removed from the initial stock to obtain the required form. The procedure can be viable for 'one-off', small or medium production runs, depending on the size, complexity and material required. These techniques may be used where the non-orthogonal geometry of the component makes manual control unfeasible, but are equally as applicable to simple parts, which may be machined from a length of standard section (Fig. 21.4).

2. *Computer controlled fabrication of tooling.* Certain other processes, such as casting and extrusion, will by their nature always require the fabrication of specific tools and moulds as part of the subsequent manufacturing process. In these instances computer-aided manufacturing is used not to fabricate the part directly (as above), but instead to create the required mould or die. CAD/CAM technology allows the fabrication of this product specific tooling either using subtractive machining processes such as CNC machining or solid freeform fabrication techniques (rapid prototyping) (Fig. 21.5).

3. *Rapid tooling.* Solid freeform fabrication (rapid prototyping) techniques allow the fabrication of components in an additive manner under direct computer control. Although solid freeform fabrication models were initially used as prototypes and visualisation tools, there is a growing field in their use to fabricate tooling, either directly or indirectly. Rapid manufacturing seeks to use SFF to revolutionise many mould and die making operations that have remained little changed for decades. Work of this nature is illustrative of an underlying trend within all forms of design engineering to develop prototyping technologies into direct and indirect production techniques wherever possible (Fig. 21.6).

22 'The non-Euclidean skin'

Contributor: Stuart Dodd

Developments in sheet metal forming

Traditionally, the pressing of many sheet metal parts required both 'male' and 'female' dies, between which the part material could be formed under pressure. CNC machining has proved an invaluable means to realise complex dies more quickly and economically than previously, but the computer control of manufacturing processes has also suggested practical and competitive applications for other methods of sheet metal forming, some of which require little or no product specific tooling. These techniques promise to bring about the realisation of many 'non-Euclidean' sheet metal parts for large-scale applications, even in small batch sizes.

Shot peen forming

Shot peen forming is a metal forming process used to create curved parts from sheet material. The process completely obviates the need for rigid dies, as the forming tool is a stream of metal shot, glass or ceramic beads. These are propelled towards the work-piece pneumatically or mechanically, and strike the sheet surface with sufficient velocity to cause its surface to deform plastically. This deformation sets up a residual stress distribution through the sheet thickness, which is resisted by the bending of the sheet towards the peening stream. This phenomenon can be used to create defined curvatures in large structural components by directing the peening stream in a number of directional axes under computer control. The process has proved commercially viable despite the fact that it is little understood and is considered something of a 'black art' within industry. In many instances the code for shot peen forming programmes must be developed using trial and error, as the deformation of the material is not completely predictable in each case.

Shot peen forming is currently utilised primarily in the aerospace industry for shaping of large sections such as wing fronts and nose cones. Shell components of this kind are already being used for the outer skin of the airbus A310 and as a structural tank for the ARIANE 4 European rocket. These parts currently remain the main application of shot peen forming together with general complex shaped and patterned panels, prototype car body panels, honeycomb panels and large tubular shapes. More recent developments concern multi-curvature geometries with differing sheet thickness. Steel, aluminium, titanium alloys and copper may all be formed.

The process would almost certainly find wider applications if it were better known and understood. Its 'die-less' nature eliminates both the high initial costs needed to produce and set up a die set, which are prohibitive for small batch production. The machinery it requires is relatively inexpensive, and its incremental nature means that parts can be inspected during the forming process. At present the largest shot peening plant can form components with maximum dimensions of 5000 × 2400 × 1000 mm and can control the direction of the peening stream in up to seven axes, suggesting many possibilities for numerically controlled panel sizes and curvatures in large scale applications.

Fluid forming

In fluid forming (also known as flex forming) sheet metal is formed over a single, rigid shape-defining tool by a flexible rubber diaphragm supported by high hydraulic pressure. What makes fluid forming competitive with traditional processes is the fact that it only requires one rigid tool half. This may be a male block or cavity die, either of which can be quickly manufactured from inexpensive materials. The use of a single rigid tool results in short lead times and lower tool cost, whilst simplifying tool modification after component design changes. These factors make the process ideal for low volume production of large sheet-metal parts.

The high uniformly distributed pressure applied by the process widens the scope of sheet metal forming. The flexible diaphragm forms scratch free parts exhibiting complex curvature, including undercuts, with different sheet thickness in all materials.

Although fluid forming is not in itself reliant on computer control, it is normally utilised in a flexible manufacturing environment using a range of CNC applications. A typical production scenario may include the computer-aided manufacturing of the rigid tool, either directly on a CNC machining centre or, perhaps, by casting from a model. Once formed the part is often trimmed to size using CNC laser cutting techniques.

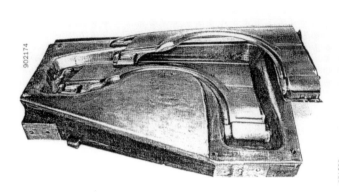
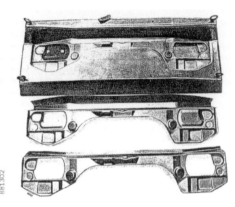

22.1
Fluid forming: tooling and parts
for car body panels.
Credit: Dr Stuart Dodd.

23 Rapid tooling

Introduction

Since the introduction of the stereolithography process designers have sought to utilise solid freeform fabrication (SFF) techniques to hasten existing paths to production. Whilst the ability to visualise physical models quickly and accurately was appreciated by designers from many disciplines, large-scale take-up of the technology is becoming increasingly motivated by the opportunity to transform the existing fabrication methods of production tooling. Although these developments have been influenced primarily by productivity concerns, rapid manufacturing, or rapid tooling as it has become known, also supports the manufacture of designs using production processes previously reserved for very high volumes. The ease with which complex geometry can be realised by these techniques has brought about several innovations in casting and mould making processes, making considerable use of powder metallurgy to transform the physical properties of the original prototypes.

Injection moulding and die casting

Existing methods

The production of moulds and dies for casting processes often has to utilise a range of machining processes for any single application. Many cast parts have complex geometry and typically display a variety of additional features such as grooves, slots and holes. Metal die casting and plastic injection moulding are well suited to creating parts of this nature, but the initial creation of the moulds themselves has previously been through the use of mainly subtractive machining processes. CNC machining techniques, particularly electrical discharge machining (EDM), have been widely used to increase the productivity of the tool making

Rapid tooling: Plastic injection moulding dies, produced from stereolithography prototypes using the 3D Keltool® process. The stereolithography station is situated in the background. Credit: 3D Systems Europe Ltd.

23.1
Prototypes and mould inserts for a plastic injection moulded valve. At least one million items can be produced from a single mould. Credit: 3D Systems Europe Ltd.

process but, despite their advantages, the process has often remained a bottleneck in the production path.

The 3D Keltool® process
The Keltool process has been developed to create production tooling directly from stereolithography (SLA®) models, reducing the lead-time and cost of moulds for plastic injection moulding and metal die casting applications. The process utilises powder metallurgy to form hard metal moulds through a series of casting operations.

Process description
1. Using CAD data SLA models are built of the mould cavities and cores, which together will make up the complete mould form.
2. These master patterns are then used to create RTV (room temperature vulcanised) silicone rubber moulds.
3. Once the master patterns are removed, the RTV moulds are filled with a mixture of A6 tooling steel powder, tungsten carbide powder and an epoxy binder. These form a slurry before chemically curing.
4. Once cured the 'green' part is removed from the mould and is sintered, by placing it within a hydrogen reduction furnace. This results in the removal of the epoxy binder.
5. This leaves a 'brown' part that is a mixture of tool steel, tungsten carbide and voids of air (formed from the burning of the binder).
6. Finally, the remaining voids are infiltrated with copper, forming a fully dense mould or insert.

The finished mould has excellent mechanical properties; the tool steel provides wear resistance and low distortion, whilst the tungsten carbide ceramic provides both hardness and durability. Additional strength and conductivity is provided by the copper. The high thermal and electrical conductivity of the mould is beneficial as it aids cooling of the mould in use, which reduces overall 'cycle' times once in production. The conductivity also supports a further application, as it enables Keltool parts to be used as shaped electrodes in EDM machining applications.

Keltool moulds can be produced in a single week, and can subsequently be used to produce large quantities of parts. For injection moulding applications millions of parts have been produced from a single mould without detrimental wear. In practice, the quantity required will dictate the subsequent production method. For example, if producing metal die casts of just a few hundred or so the casting process may be performed manually, but if thousands were required the mould may be incorporated into a mechanised casting facility.

At present, the maximum size of part is 150 × 215 × 100 mm, but additional moulds may be keyed together for larger applications. If required, moulds can be further processed using subtractive metal machining techniques, either to add additional features or to improve surface finish.

Mould and die fabrication using selective laser sintering®

Similar applications are supported using the DTM Sinterstation®. In fact, the selective laser sintering (SLS®) process is able to produce certain types of mould directly, by fabricating with materials developed specifically for the task. Copper polyamide is used as a build material to produce moulds for plastic injection moulding applications, which are each capable of producing several hundred parts. For more durable moulds, including metal die casting applications, a multi-stage process is used. The mould is first built using a polymer coated tool steel to produce the 'green' part. Subsequent heating in a furnace and infiltration with bronze produces a durable mould capable of producing several hundred metal casts from either aluminium, magnesium or zinc. These moulds also enable the production of high volumes of plastic parts.

Wax pattern production for 'lost wax' investment casting

Solid freeform fabrication techniques can also be used in conjunction with existing techniques, which themselves remain unchanged despite developments in pattern and mould creation.

The investment casting process uses expendable wax patterns as the basis for the formation of a hollow ceramic shell into which metal is poured. This process is economical in batches from as low as a dozen up to those of several thousand.

Unless a single casting is required, additional identical wax patterns must be produced to obtain a batch of any size. This has traditionally been achieved by making either a rubber mould or an aluminium die, from which further wax patterns may be cast.

SFF techniques have modified this process in a number of ways. Rapid prototypes can now be built directly from a

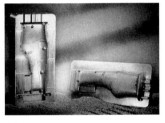

23.2
Plastic injection moulding dies.
Credit: DTM Corporation.

wax-like substance, which can be 'burnt out' in the traditional manner. This may be appropriate for unique castings or for small batch sizes. For higher volume applications SFF techniques may be employed to create the die itself. This takes place in a manner similar to that described for injection moulding. In both cases the nature of the investment casting process itself remains unchanged. In this manner existing knowledge, expertise and plant associated with the foundry remain relevant.

Sand casting cores and moulds

Emerging techniques and ancient crafts are also combined when SFF is used with the process of sand casting. Any object, which is sufficiently rigid and within certain geometric constraints, can be used as a pattern for sand casting. As a result virtually all SFF processes can be used to build patterns, providing it can be removed from the sand mould without disturbing its imprint. Although this use shortens significantly pattern production time, the traditional constraints placed upon form by the mould making process still remain. For this reason polymer coated foundry sands are now used to build SFF moulds directly. As the mould is built layer by layer, and a pattern does not have to be removed from the cavity, part geometry can be more complex. Prototype ferrous and aluminium castings can be produced in this manner, exhibiting forms that would be impossible using a pattern and core box (Fig. 21.6). This technique is a feature of the SLS process.

Sinterstation and SLS are trademarks of the DTM Corporation.

3D Keltool and SLA are trademarks of 3D Systems.

Select bibliography

Ayres, R.U. (1990). *Computer Integrated Manufacturing*. Vol. I. London: Chapman & Hall.

Ayres, R.U., Haywood, W. and Tchijov, I. (eds.) (1992). *Computer Integrated Manufacturing*. Vol. III. London: Chapman & Hall.

Babbage, C. (1835). *On the Economy of Machinery and Manufactures*. London: Charles Knight.

Bagnasco, A. and Sabel, C. (eds.) (1995). *Small and Medium-Size Enterprises*. London: Pinter.

Barthes, R. (1957). *Mythologies*. London: Vintage.

Bassalla, G. (1988). *The Evolution of Technology*. Cambridge: Cambridge University Press.

Bijker, W.E. (1995). *Of Bicycles, Bakelites, and Bulbs*. Cambridge, MA: MIT Press.

Bonsiepe, G. (1994). *Interface: An Approach to Design*. Maastricht: Jan van Eyck Akademie.

Burden, W.W. (1944). *Broaches and Broaching*. New York: Broaching Tool Institute.

Burton, A. (1994). *The Rise and Fall of British Shipbuilding*. London: Constable.

Campbell-Kelly, M. (ed.) (1994). *Charles Babbage: Passages from the Life of a Philosopher*. New Brunswick, NJ: Rutgers University Press.

Day, A. (1997). *Digital Building*. Oxford: Laxton's.

Donald, A. (ed.) (1995). *World War II and the American Dream*. Cambridge, MA: MIT Press.

Dormer, P. (1994). *The Art of the Maker*. London: Thames & Hudson.

Evans, R. (1997). *Translation from Drawing into Building and Other Essays*. London: Architectural Association.

Ferguson, E.S. (1992). *Engineering and the Mind's Eye*. Cambridge, MA: MIT Press.

Giedion, S. (1954). *Walter Gropius: Work and Teamwork*. London: The Architectural Press.

Grimshaw, N. (1998). *Fusion*. London: Grimshaw & Partners.

Groák, S. (1992). *The Idea of Building*. London: E. & F.N. Spon.

Groák, S. (1983). Building Processes and Technological Choice. *Habitat International*, Vol. 7, No. 5/6.

Groák, S. and Ive, G. (1986). Economics and Technological Change: Some Implications for the Study of the Building Industry. *Habitat International*, Vol. 10, No. 4.

Gropius, W. (1935). *The New Architecture and the Bauhaus*. London: Faber and Faber.

Hard, M. and Jamison, A. (eds.) (1998). *The Intellectual Appropriation of Technology*. Cambridge, MA: MIT Press.

Hartoonian, G. (1994). *Ontology of Construction*. Cambridge: Cambridge University Press.

Haugeland, J. (1985). *Artificial Intelligence*. Cambridge, MA: MIT Press.

Henderson, K. (1999). *On Line and On Paper*. Cambridge, MA: MIT Press.

Hill, J. (ed.) (1998). *Occupying Architecture: Between the Architect and the User*. London: Routledge.

Horner, J.G. (1950). *Pattern-Making for Engineers*. London: The Technical Press.

Hyman, A. (1982). *Charles Babbage: Pioneer of the Computer*. Oxford: Oxford University Press.

Hyman, A. (ed.) (1989). *Science and Reform*. Cambridge: Cambridge University Press.

Jones, B. (1997). *Forcing the Factory of the Future*. Cambridge: Cambridge University Press.

Judge, A.W. (ed.) (1950). *Toolroom Practice*. Vol. II. London: Caxton Publishing Company.

Kalpakjian, S. (1995). *Manufacturing Engineering and Technology*. Reading, MA: Addison-Wesley.

Kochan, D. (ed.) (1986). *CAM Developments in Computer-Integrated Manufacturing*. Berlin: Springer Verlag.

Lukin, J. (1885). *Turning for Amateurs*. London: Upcott Gill.

MacKenzie, D. and Wajcman, J. (eds.) (1985). *The Social Shaping of Technology*. Milton Keynes: Open University Press.

McCullough, M. (1996). *Abstracting Craft*. Cambridge, MA: MIT Press.

McLuhan, M. (1964). *Understanding Media*. Cambridge, MA: MIT Press.

Petroski, H. (1993). *The Evolution of Useful Things*. London: Pavilion Books.

Potter, N. (1990). *Models and Constructs*. London: Hyphen Press.

Pursell Jr, C.W. (ed.) (1981). *Technology in America*. Cambridge, MA: MIT Press.

Pye, D. (1968). *The Nature and Art of Workmanship*. Cambridge: Cambridge University Press.

Pye, D. (1978). *The Nature and Aesthetics of Design*. London: The Herbert Press.

Rice, P. (1994). *An Engineer Imagines*. London: Ellipsis.

Rolt, L.T.C. (1965). *Tools for the Job*. London: B.T. Batsford.

Scheer, A.-W. (1988). *CIM: Computer Integrated Manufacturing*. Berlin: Springer Verlag.

Sebestyen, G. (1998). *Construction: Craft to Industry*. London: E. & F.N. Spon.

Smith, G.T. (1993). *CNC Machining Technology*. London: Springer Verlag.

Star, S. (1995). *The Cultures of Computing*. Oxford: Blackwell Publishers.

Street, A. and Alexander, W. (1944). *Metals in the Service of Man*. London: Penguin Books.

Susnjara, K.J. (1998). *Furniture Manufacturing in the New Millennium*. Dale: Thermwood Corporation.

Susnjara, K.J. (1999). *Three Dimensional Trimming and Machining.* Dale: Thermwood Corporation.

Swade, D. (1991). *Charles Babbage and his Calculating Engines.* London: Science Museum.

Thyer, G.E. (1988). *Computer Numerical Control of Machine Tools.* Oxford: Newnes.

Tomlinson, G. (1984). *Industrial Workshop Practice.* Harlow: Longman.

Various (1959). *General Engineering Workshop Practice.* London: Odhams Press.

Various (1947). *Practical and Technical Encyclopaedia.* London: Odhams Press.

Vickers, G.W., Ly, M.H. and Oetter, R.G. (1990). *Numerically Controlled Machine Tools.* New York: Ellis Horwood.

Volti, R. (1992). *Society and Technological Change.* New York: St Martin's Press.

White, R.B. (1965). *Prefabrication: A History of its Development in Great Britain.* London: HMSO.

Willis, D. (1999). *The Emerald City.* New York: Princeton Architectural Press.

Index